World of science

THE WORLD OF MAN

BAY BOOKS LONDON & SYDNEY

1980 Published by Bay Books
157–167 Bayswater Road, Rushcutters
Bay NSW 2011 Australia

National Library of Australia
Card Number and ISBN 0 85835 278 8
Design: Sackville Design Group
Printed by Tien Wah Press, Singapore.

THE WORLD OF MAN

Man is a warm blooded creature that bears live young and produces milk on which to feed them, so he belongs to the group called mammals. But man walks upright on two legs and has very little hair on his body.

Man's brain is much more developed than animal's brains and he can think, work out problems and form abstract ideas and speak. He does not always have to be able to see and feel something before he can think about it. Abstract ideas include love, friendship and beauty, which we all understand although we cannot see them.

Man's brain and his power to speak allow him to pass on his ideas to others. Man is also unusual in the way he uses his hands, to handle tools, write and create works of rt such as paintings and sculpture. Because of these bilities, man is the only creature on Earth to have vilisation and culture, art, religion and science.

He is also the only creature to study himself and his rld. These studies have allowed man to survive in all ds of surroundings, from the polar regions to outer ce. From the time man learned to use fire, to farm the d and to understand things such as medicine, he has n able to control at least some of his world to suit nself.

This lateral section through a human brain shows its complex structure with wrinkles, folds and fissures. The brain is made up of about 30, 000, 000, 000 cells, which are linked by nerve fibres. Electrical impulses transmit information between the cells and the other parts of the human body. Man's brain has enabled him to elevate himself above the other mammals.

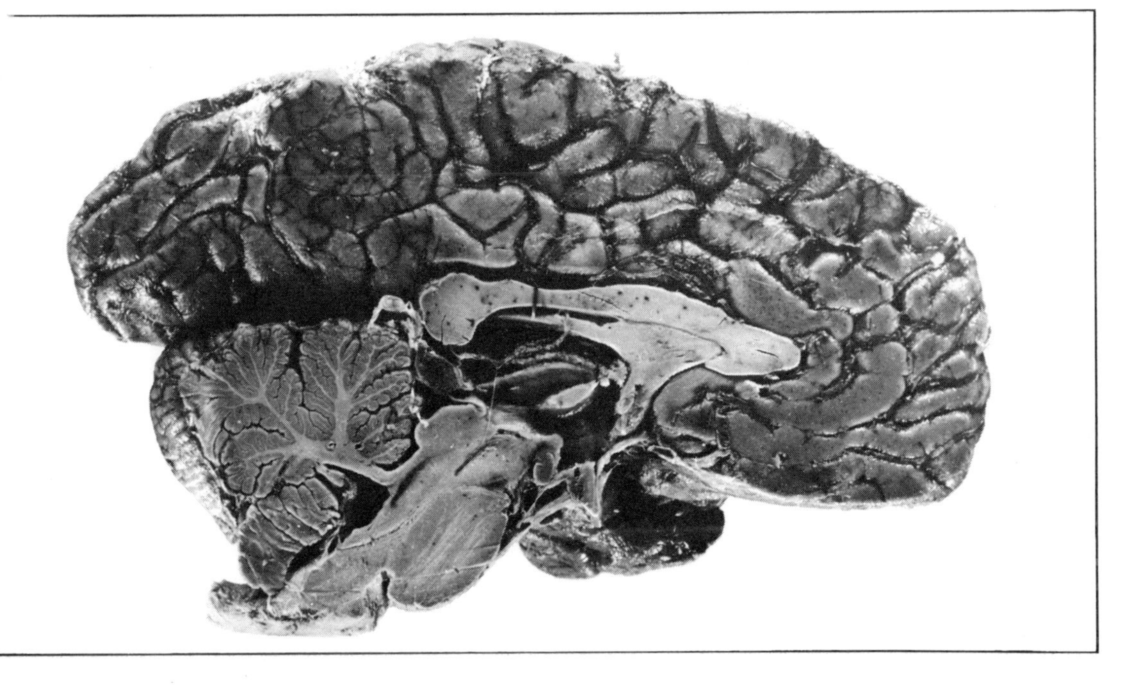

EVOLUTION AND ANTHROPOLOGY

Evolution

Man and the apes have developed from the same ancestors, although the main differences between them are man's larger brain and his ability to walk upright and speak.

The process by which modern man developed from his early ancestors is called *evolution.* Through fossils, very old bones preserved in layers of rock, we can imagine how man developed and changed from an ape-like creature to modern man, over a period of several hundred thousand years. Until recently, the oldest man-like fossil anyone had discovered was called *Australopithecus*, about 1,750,000 years old. Then, in 1972, Dr Richard Leakey found a fossil skull in Kenya that had a much larger brain than the *Australopithecus.* This fossil was about 2,600,000 years old.

Several other fossils have been found which look more like modern man. One of these, called *Pithecanthropus,* was very like us although he had jutting-out ridges over his eyes and some ape-like features. Other remains such as Java man and Peking man are called *Homo erectus.* All

Man gradually evolved from the early Australopithecus, who appeared during the Tertiary period, to Neolithic man, the direct ancestor of modern Man. After he discovered fire, Man went on to become a tool and weapon maker and cast his hunting spears and arrow heads in iron. He soon learnt to grow his own food and became a farmer as well as a hunter.

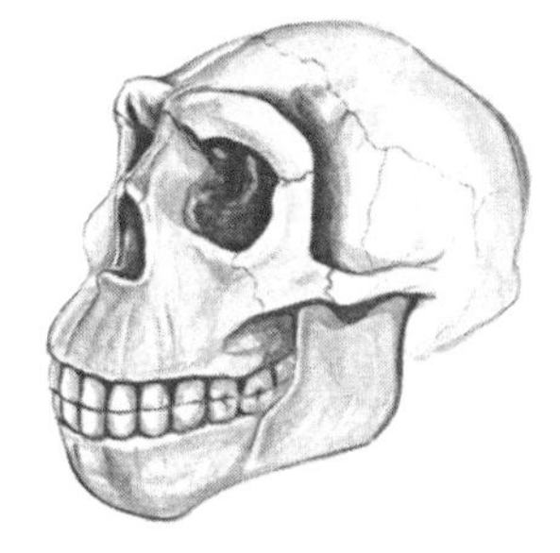

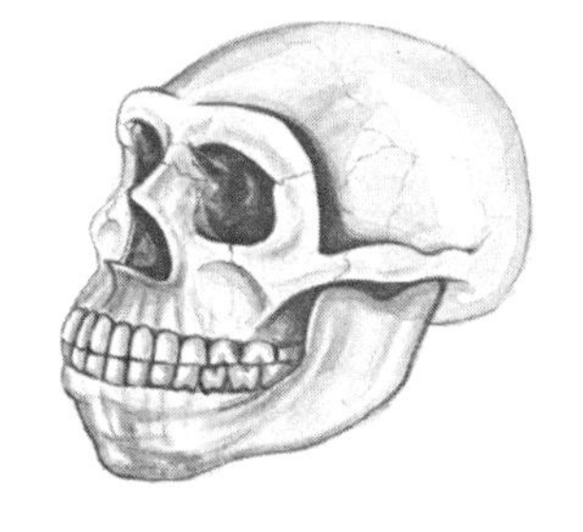

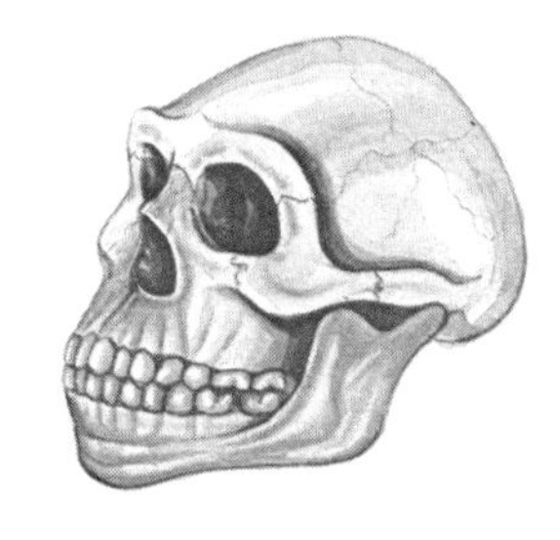

the known fossils in this group are at least 400,000 years old, and from this creature, several types of man seem to have developed. One was Neanderthal man, who looked very primitive with his thick, stocky body and jutting eye ridges, but he did have a large brain. The Neanderthals

Above: This illustration shows the skulls and facial characteristics of Rhodesian man, Homo erectus and Neanderthal man.

CAUCASOID

ENGLISH

ITALIAN

INDIAN

Caucasoid
The Englishman, Italian and Indian above are all Caucasians. This racial group is widespread in Europe, the Middle East, North Africa and India, and they have skin colours ranging from fair to dark.

used stone tools and buried their dead. They lived in Europe about 35,000 years ago, when men just like ourselves replaced them. This change happened too quickly for modern man to have developed from Neanderthal man, so he must have evolved separately. We call modern man *Homo sapiens* ('thinking man').

Some of the earliest fossils of *Homo sapiens* are over 100,000 years old. Among the most famous are the fossils found at Cro-Magnon in France. These early men lived in caves which they decorated with drawings, so they were the first known artists.

Races

All modern men are placed in the family *Homo sapiens.* However, because some people have similar features, they are often divided into groups, or races. These groups are usually: the Caucasoid (white-skinned); Negroid (black-skinned); Mongoloid (yellow-skinned and including the American Indians); Australoid (the Australian Aborigines); and Capoid (the bushmen of South Africa).

Race is often confused with *nationality,* the nation a person belongs to. But the people of any one country are often descended from many races so race is a much wider description than nationality.

CAPOID

KALAHARI BUSHMAN

Capoid
This Kalahari bushman from the deserts of southern Africa is a member of the Capoid group.

Negroid
The Ghanaian Ashanti and the Ugandan Bahima are African members of the Negroid group.

NEGROID

ASHANTI:(GHANA)

BAHIMA: (UGANDA)

AUSTRALOID

MBOTGOTE: (MALEKULA ISLAND)

FIJIAN

ABORIGINE

MONGOLOID

ESKIMO

NAGA:
(INDO-BURMESE BORDER)

CHINESE

Anthropology

The study of man is called *anthropology.* There are two main areas which anthropologists study. One is *physical anthropology,* which is concerned with the appearance of the human body. These scientists study fossils to try to discover how man evolved. They also study the many races of man alive today. By looking at skin colours, head and body shape, hair colour and waviness, and many other things, they may be able to find links between people in different parts of the world.

The other branch is called *cultural anthropology,* because it deals with man's activities rather than his looks. These activities include family life, tribal and community life, marriage customs, bringing up children, religious and magical beliefs, agriculture, medicine and the use of tools.

An important area of cultural anthropology is the study of language, which can tell us a lot about the history and culture of any people and show how they are linked with other peoples, often in different parts of the world.

Australoid
Although they have different facial characteristics and features, the Mbotgote man from Malekula Island, the Fijian and the Australian Aborigine are all Australoid peoples with dark skins.

Mongoloid
With their yellowish skin, dark brown eyes and straight black hair, the Eskimo, Burmese man and Chinese are all Mongoloids.

Cells

This cut-away view of the human cell shows its complex structure. The nucleus is the centre of the cell and contains gene-carrying chromosomes. Cells are the basic living units of the body and carry out all of the fundamental processes of life, from forming tissues and fighting germs to replacing enzymes.

The human body is made up of many millions of tiny units called *cells,* each of which works by itself and with other cells to carry on the process of life.

A human body functions because of the way the various parts work together in an organised way. The cells are grouped together into various types of *tissues.* These tissues are built into parts of the body called *organs* and these work together in *organ systems.*

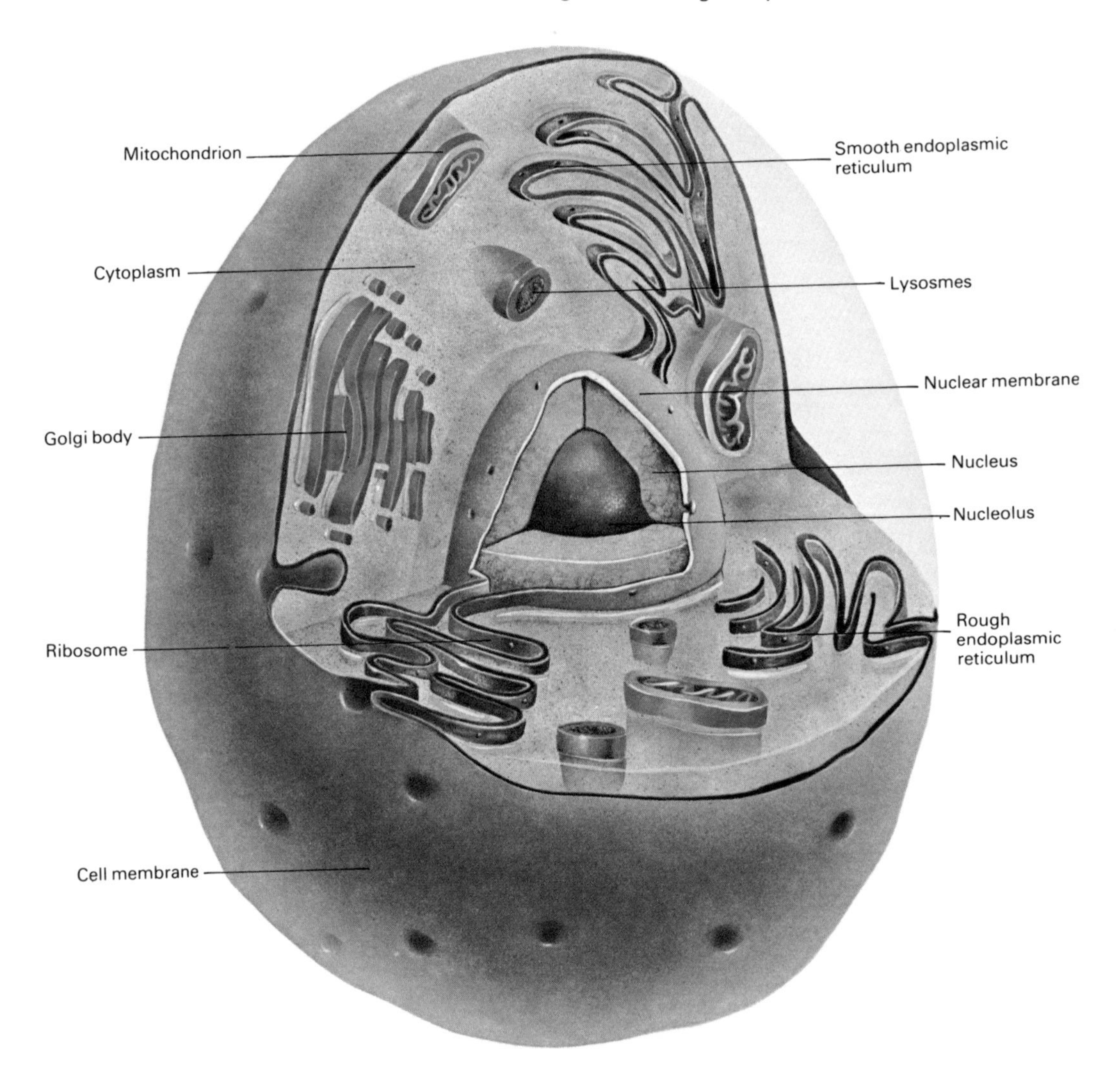

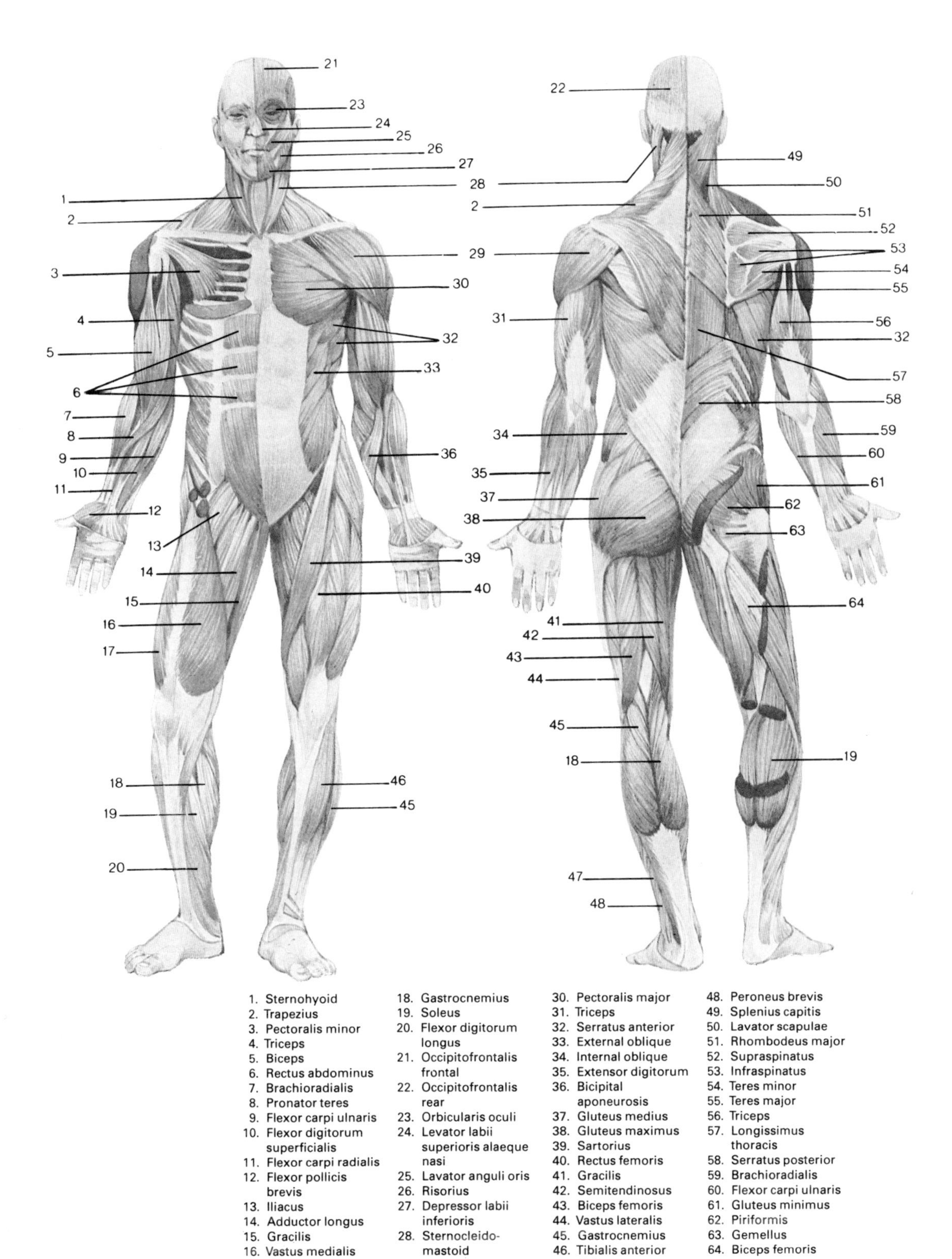
1. Sternohyoid
2. Trapezius
3. Pectoralis minor
4. Triceps
5. Biceps
6. Rectus abdominus
7. Brachioradialis
8. Pronator teres
9. Flexor carpi ulnaris
10. Flexor digitorum superficialis
11. Flexor carpi radialis
12. Flexor pollicis brevis
13. Iliacus
14. Adductor longus
15. Gracilis
16. Vastus medialis
17. Vastus lateralis
18. Gastrocnemius
19. Soleus
20. Flexor digitorum longus
21. Occipitofrontalis frontal
22. Occipitofrontalis rear
23. Orbicularis oculi
24. Levator labii superioris alaeque nasi
25. Lavator anguli oris
26. Risorius
27. Depressor labii inferioris
28. Sternocleido-mastoid
29. Deltoid
30. Pectoralis major
31. Triceps
32. Serratus anterior
33. External oblique
34. Internal oblique
35. Extensor digitorum
36. Bicipital aponeurosis
37. Gluteus medius
38. Gluteus maximus
39. Sartorius
40. Rectus femoris
41. Gracilis
42. Semitendinosus
43. Biceps femoris
44. Vastus lateralis
45. Gastrocnemius
46. Tibialis anterior
47. Soleus
48. Peroneus brevis
49. Splenius capitis
50. Lavator scapulae
51. Rhombodeus major
52. Supraspinatus
53. Infraspinatus
54. Teres minor
55. Teres major
56. Triceps
57. Longissimus thoracis
58. Serratus posterior
59. Brachioradialis
60. Flexor carpi ulnaris
61. Gluteus minimus
62. Piriformis
63. Gemellus
64. Biceps femoris

Bones and muscles

The bone framework which supports your body is called the *skeleton.* Apart from giving shape, the skeleton supports the system of muscles which enables us to move. The muscles operate the bony system at its joints to allow movement to take place. The bones also protect the softer organs of the body such as the heart and lungs, which have a shelter of bones called the rib cage. The hard bony skull protects the brain from harm. A human baby is born with about 305 individual bones in its skeleton but many of these later grow together, so that an adult skeleton usually has 206 bones.

To control movement in the body, messages must come from the brain to the muscles. These messages are sent down a thick nerve stem called the *spinal cord,* which extends from the brain down the backbone or *vertebral column.* Pairs of spinal nerves branch off from each side of this cord to relay messages to different parts of the body.

Attached to the skeleton at various points are strong meaty tissues called muscles. When a muscle receives a signal from your brain, it *contracts* to pull on a bone or some other part of the body, causing movement. There are more than 600 muscles in the human body and they make up about 40 per cent of its weight. Every single movement from walking, writing and breathing to blinking an eye or digesting food is caused by the action of muscles.

This cross-section through the muscle shows that it is made up of a bundle of fibres. Muscles account for about 35 to 45 per cent of the weight of the human body. They can be flexed or extended as is shown here by the movement of the arm. There are three types of muscles — skeletal, cardiac and smooth — of which skeletal, which cover the skeleton, are the largest and most powerful.

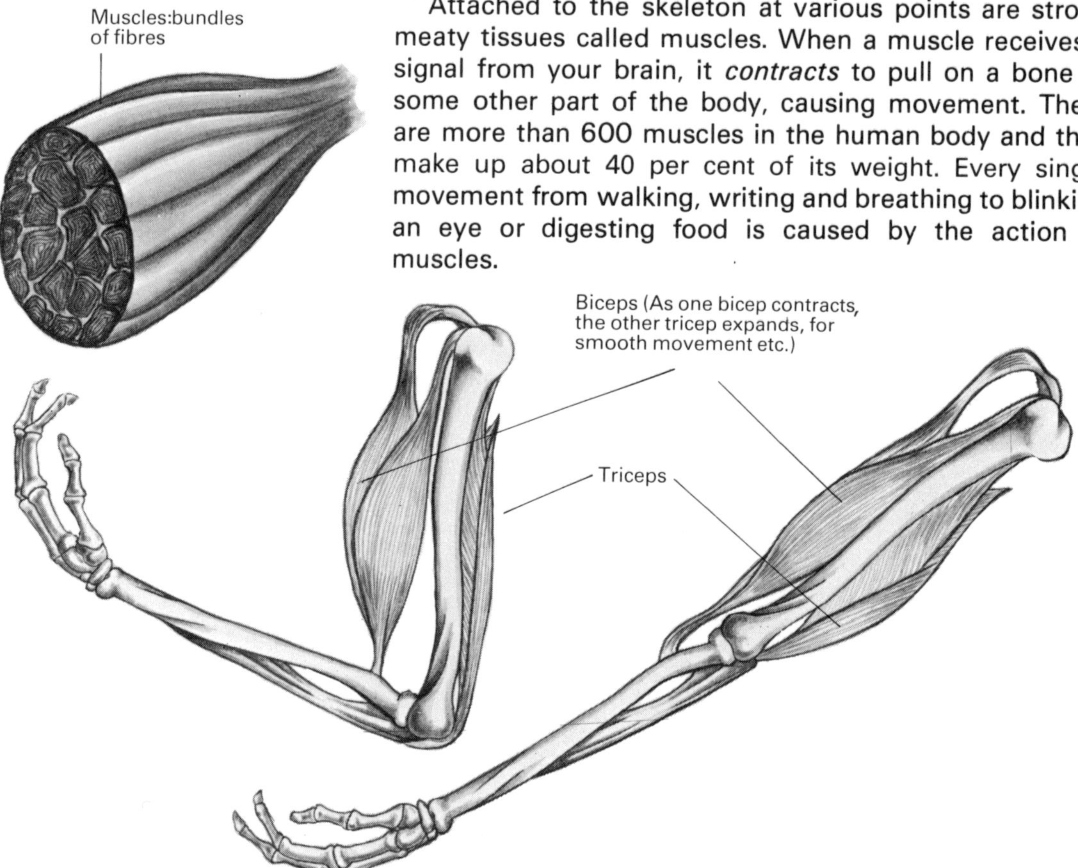

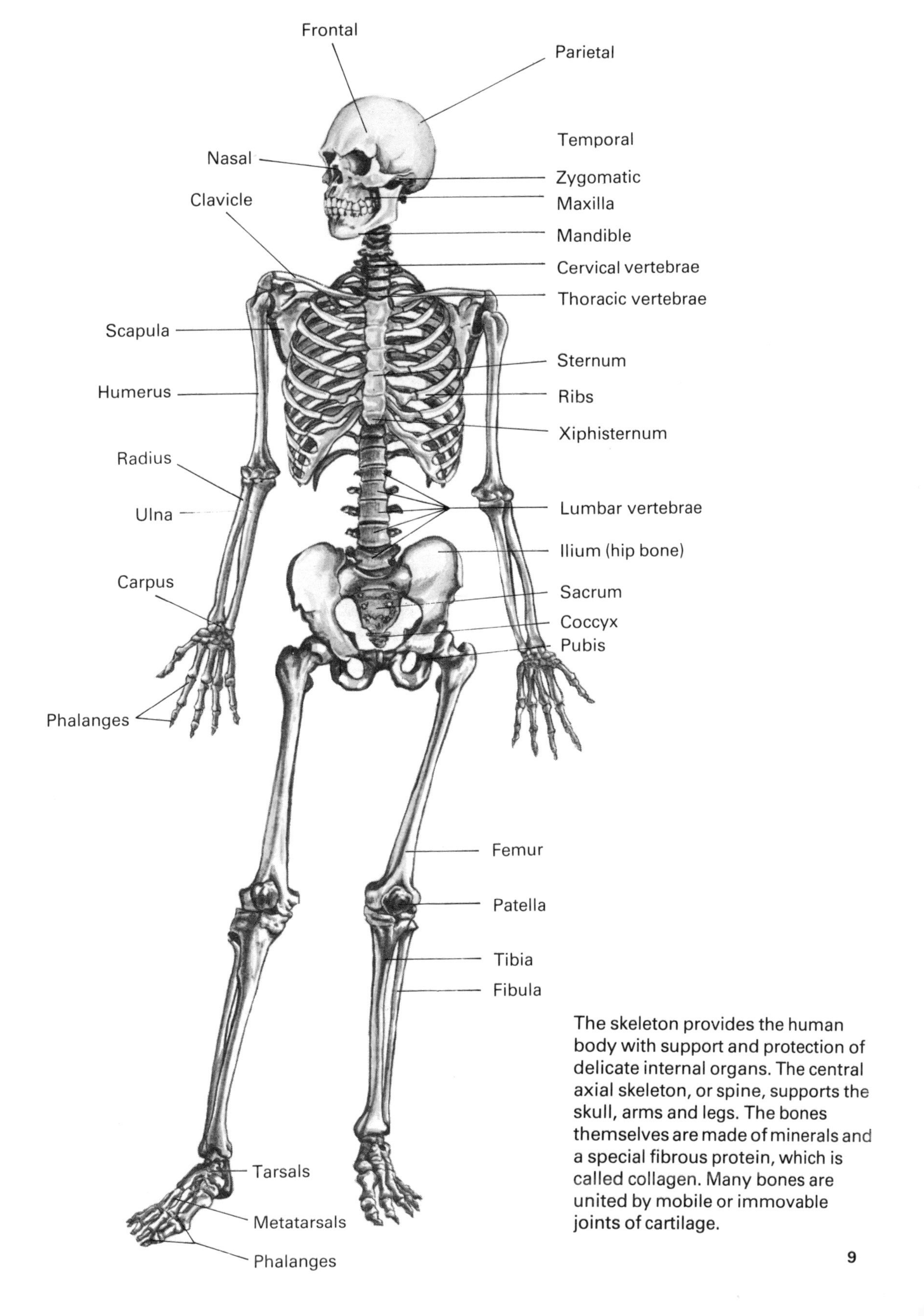

The skeleton provides the human body with support and protection of delicate internal organs. The central axial skeleton, or spine, supports the skull, arms and legs. The bones themselves are made of minerals and a special fibrous protein, which is called collagen. Many bones are united by mobile or immovable joints of cartilage.

How muscles work

There are two kinds of muscles: those under the control of the brain, called *voluntary muscles,* and those which cannot normally be controlled, called *involuntary muscles.* Voluntary muscles cause movements such as walking or waving an arm. The involuntary muscles control things like heartbeat and blood pressure.

When a nerve signal reaches a muscle it makes it contract, or shorten. Most muscles pull on a bone, causing a swivelling movement at a joint. Muscles cannot push, so for each muscle, there is a structure called an *antagonist* muscle which pulls the bone back into its original position. The signals from the brain ensure that these movements are smooth and controlled, rather than jerky and violent.

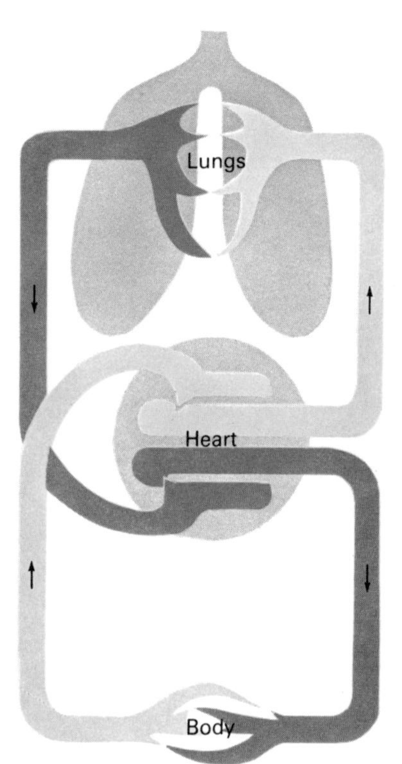

This representation of the circulatory system shows how blood is pumped by the heart around the human body. It is carried away from the heart by the arteries and returned via the veins. The blood is oxygenated in the lungs and then carried to the heart by the veins. It is then pumped from the heart in the arteries, then passes through a network of capillaries where the oxygen is exchanged for carbon dioxide and waste. The deoxygenated blood is transported by the veins back into the heart before being reoxygenated in the lungs.

The blood

The system of organs which takes blood to all parts of the body is known as the *circulatory system.* In man and other advanced creatures, this consists of the heart and various kinds of tubelike organs called *blood vessels.* Blood is passing through these tubes all the time in a never-ending cycle.

Man's heart has four sections called *chambers* which are really like two separate pumps. The right side pumps blood to the lungs, the left side pumps blood to the rest of the body.

There are almost 100,000 kilometres of blood vessels in the human body. These vessels are divided into the *arteries* which carry the blood away from the heart to parts of the body and the *veins,* which carry the blood back to the heart again.

Blood is the body's transport medium. It carries food and oxygen to all the cells, and carries away carbon dioxide and other wastes. It transports heat, keeping the body at a constant temperature. Blood also contains special cells called *antibodies,* which fight infection.

As well as blood vessels, there is another network of similar vessels which contain a watery fluid called lymph. This system contains lymph nodes that act as filters to remove harmful bacteria. The tonsils in your throat are part of this system, which produces antibodies to fight infection.

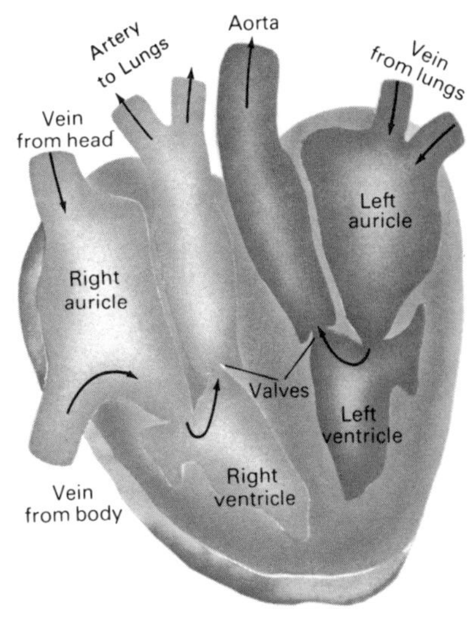

Above: This cross-section of the heart shows its four chambers: the ventricles and the atria. The blood passes from the venae cavae into the right auricle, and the tricuspid valve controls its rate of flow into the right ventricle. At the same time, the left auricle fills with blood and the valve to the left ventricle is forced open under pressure. When the distended ventricles are full of blood, the valves snap shut and the blood passes out into the aorta or the pulmonary artery. Both ventricles eject roughly equal quantities of blood as they pump it out simultaneously.

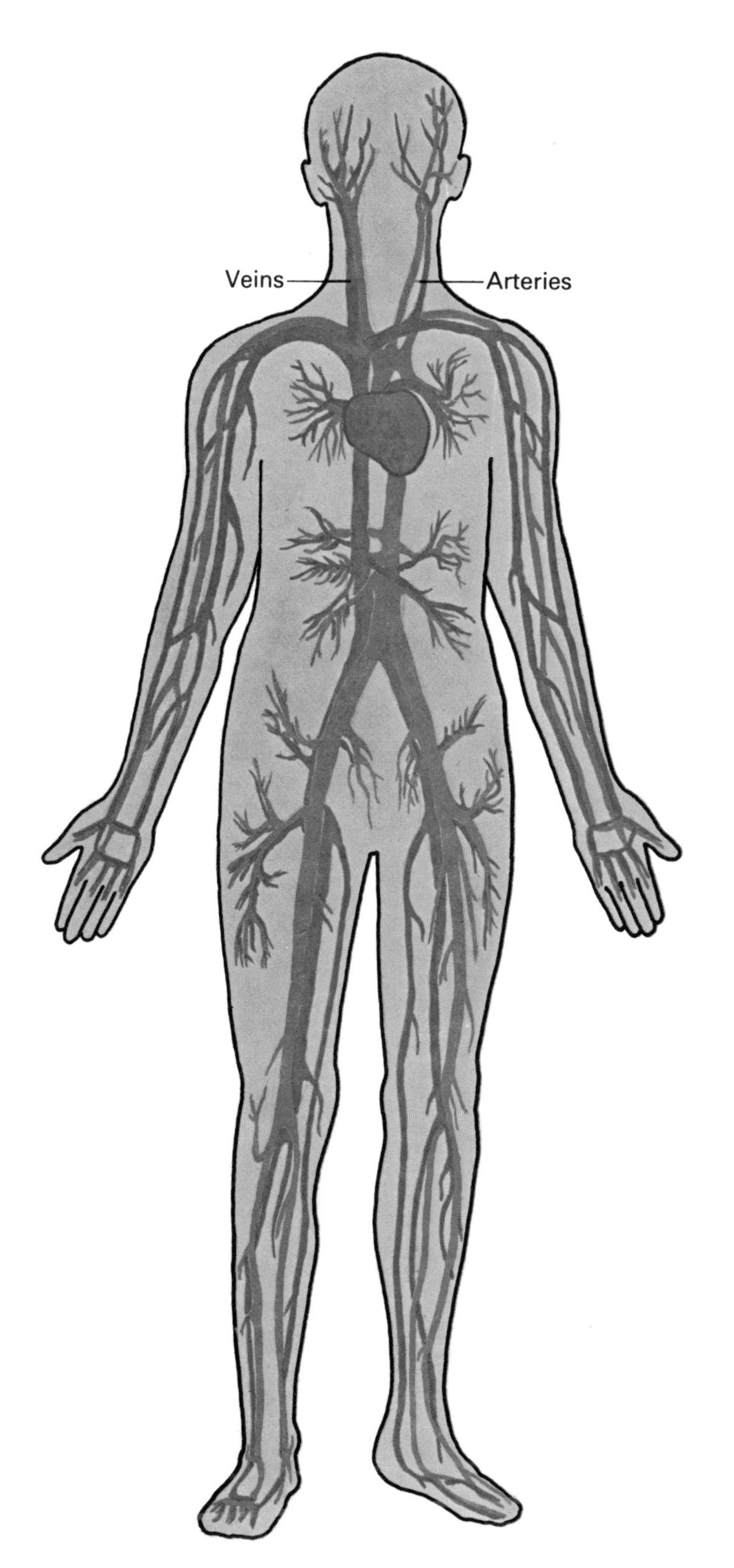

The circulatory system is the means by which blood is transported around the body. There is an astonishing 96, 500 km of capillaries that make up this complex transportation network. The circulating blood carries nutrients and oxygen to the body cells, removing waste and replenishing tissues.

Respiration

When air is breathed into the body, it passes through the trachea into the bronchi and then into the lungs where an exchange of gases takes place. Carbon dioxide is expelled from the body and the fresh reoxygenated blood is pumped via the heart to the body.

This is usually called breathing, and is one of the basic processes of life. The action of breathing moves air in and out of the lungs. The lungs are large, spongy, flexible organs in the chest cavity. When this cavity expands or contracts, becoming larger or smaller, the lungs expand or contract as well. This makes air flow in and out through the nose or mouth and the tube connecting the throat to the lungs, called the *trachea.* Two things make the chest cavity expand. A sheet of muscle called the *diaphragm* at

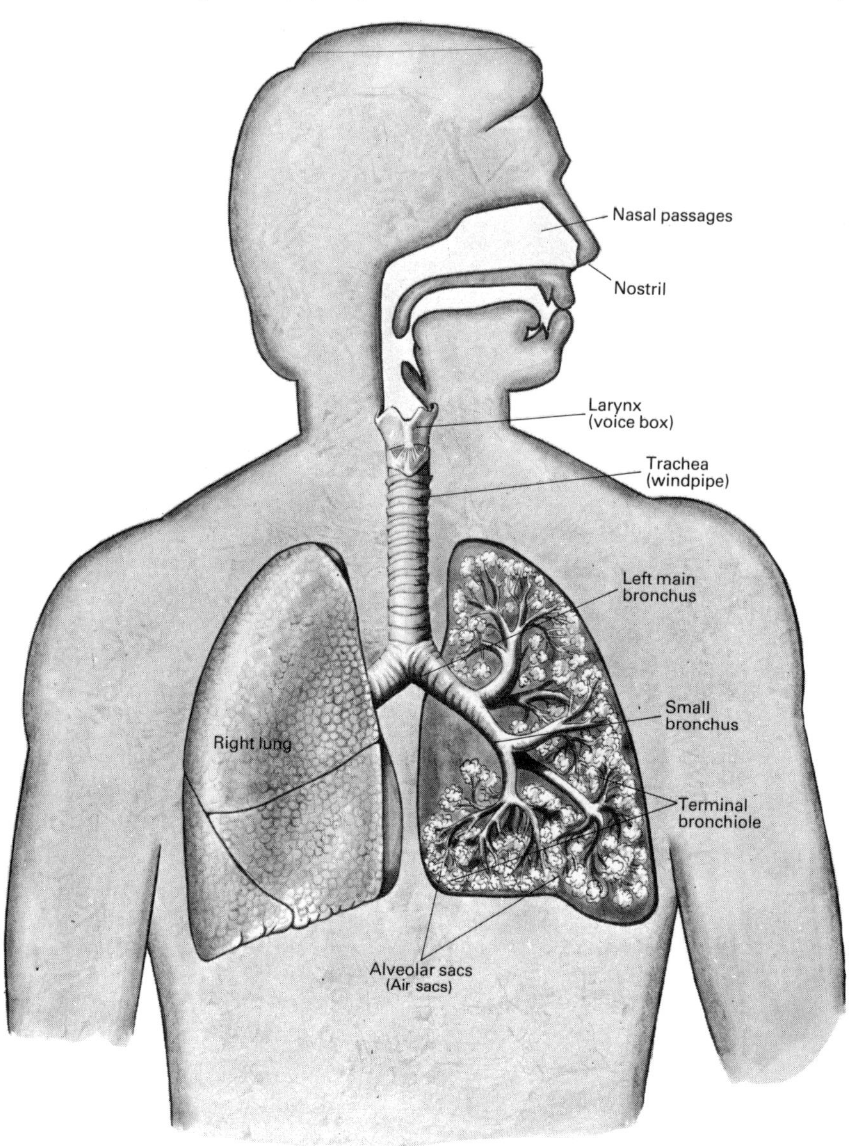

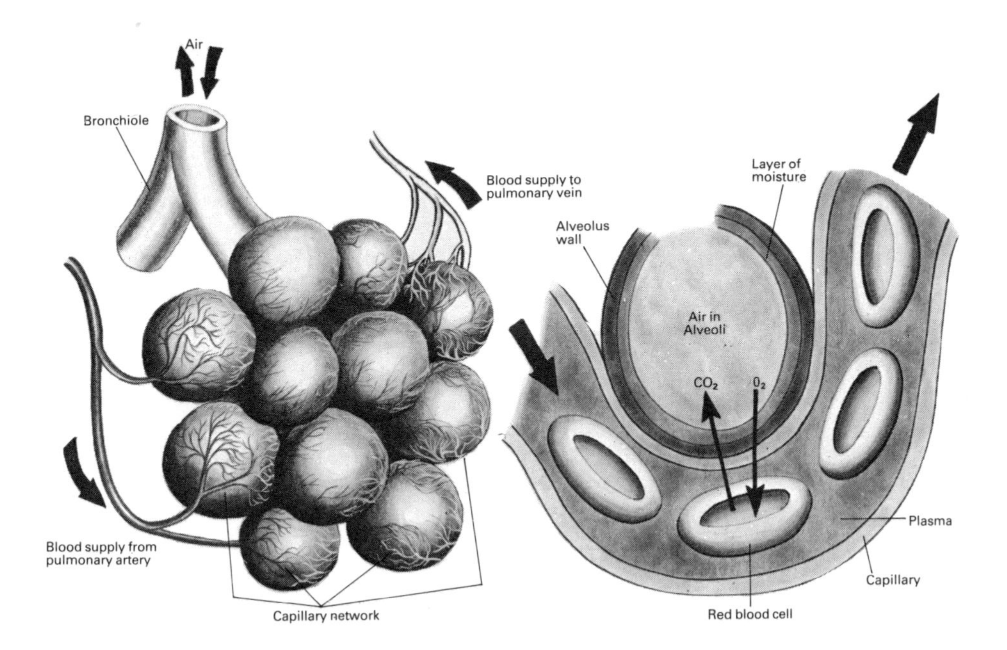

the bottom of the chest cavity is domed upwards when relaxed, but contracts and flattens out when, at the same time, the muscles between the ribs also contract and this causes the rib cage to expand. This movement corresponds to the rhythm of the breathing, which is controlled by a part of the brain called the *respiratory centre*. You cannot stop breathing completely even if you want to, for it is only a short while before this centre takes over and forces you to breathe.

Human beings have two lungs, each roughly cone shaped. The left lung is smaller than the right because some space in the chest cavity or *thorax* is taken up by the heart.

The trachea, or windpipe, brings the air you breathe down from the back of the throat. High in the chest, the trachea divides into two small tubes called *bronchi*, which connect with each lung. The bronchi divide again and again into a network of smaller tubes known as the bronchial tree. The smallest tubes of all are called *bronchioles*.

From each bronchiole open small air sacs which look like miniature bunches of grapes. The walls of the air sacs are folded to form the smallest cavities of all, the tiny *alveoli*. There are so many alveoli, more than 600 million in the two lungs, that if they were smoothed out they would have a total surface of about 230 square metres. Around the alveoli is a network of the small blood vessels called *capillaries*, so that the alveoli are like lots of tiny bubbles of air in a sea of blood. All this makes it easy for

Inside the grape-like clusters of the alveoli, stale blood is reoxygenated. The alveoli are minute air pockets which are surrounded by blood vessels. Oxygen is filtered out through their walls into the blood vessels to be carried away around the body. At the same time, carbon dioxide waste is expelled from the blood vessels into the alveoli.

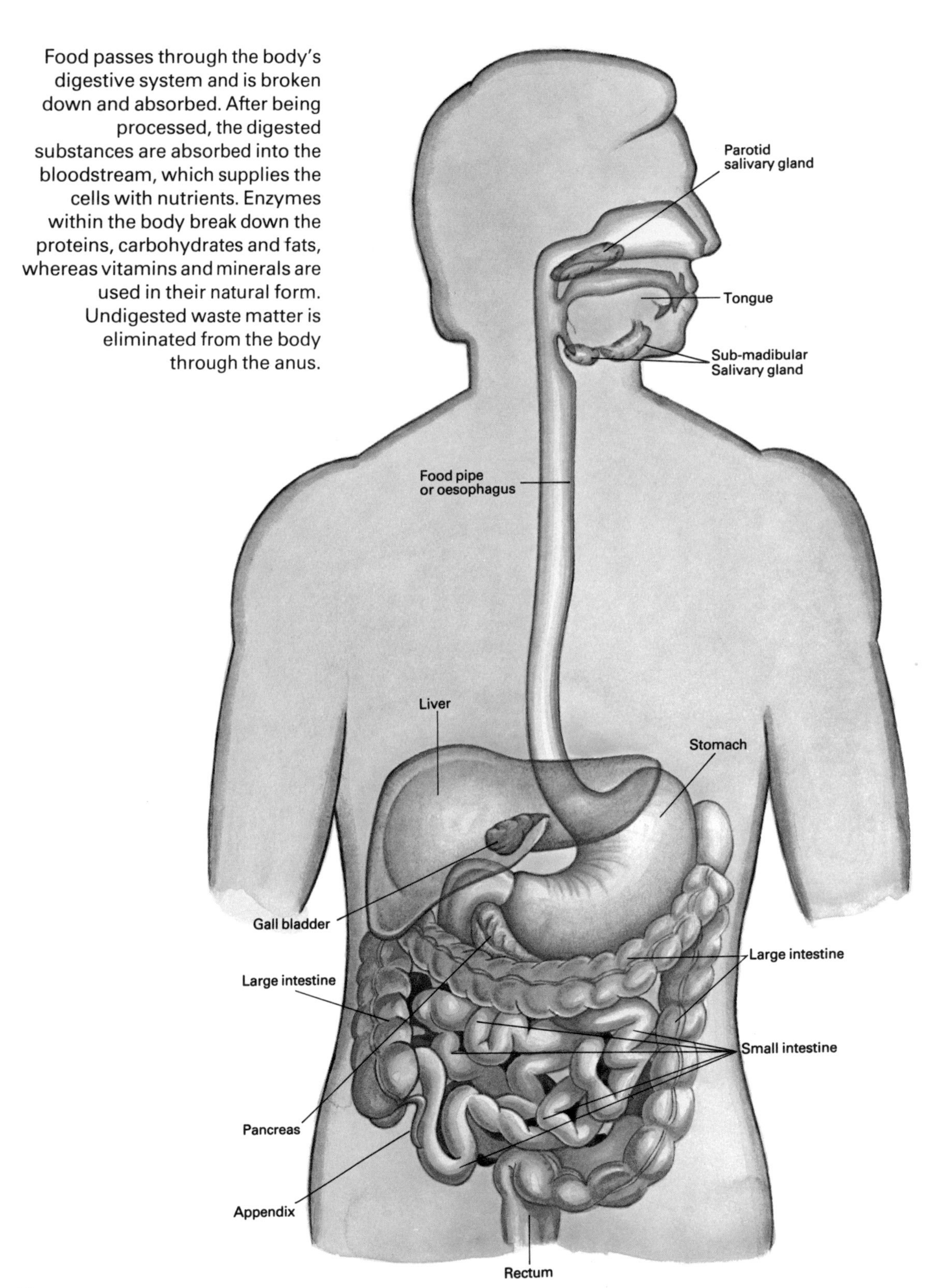

Food passes through the body's digestive system and is broken down and absorbed. After being processed, the digested substances are absorbed into the bloodstream, which supplies the cells with nutrients. Enzymes within the body break down the proteins, carbohydrates and fats, whereas vitamins and minerals are used in their natural form. Undigested waste matter is eliminated from the body through the anus.

gases to pass between the air and the blood. Blood which contains little oxygen reaches the heart, passes to the lungs and stocks up on oxygen, then returns to the heart again to circulate through the body.

An adult man's lungs can hold about 11 to 12 litres of air altogether. In normal, relaxed breathing there is only about half a litre of air passing in and out. When a man is breathing heavily while running or swimming hard, for example, much more air is used.

Digestion

Food is made up of many types of complicated chemical substances. Before the body can use these they must be 'broken down' into simpler substances. Starch in bread, for example, cannot be used by the body as it is, but must be turned into simple sugars in the body's food processing factory which we call the *digestive system.*

Digestion begins when you put food into your mouth. Chewing breaks the food up and mixes it with the fluid in your mouth called *saliva* which assists the further breakdown of the food. When we swallow our food and it passes into the *oesophagus* or *gullet,* a wave-like muscular movement called *peristalsis* carries it down to the stomach. The time the food stays in the stomach depends on the type and amount of food.

The stomach muscles contract and churn the food into a half-liquid mass called *chyme.* At the same time, gastric glands in the stomach wall send out gastric juice to begin digesting the proteins in the food. Other substances in the stomach are also at work and, eventually, the food is thoroughly broken up. When this is complete, a ring of muscles called the *pyloric sphincter* relaxes and lets the chyme out of the stomach into an area called the *duodenum.* Different types of juices pour into the duodenum and continue digesting the food. The contents of the duodenum are then sent along by peristalsis to other parts of the small intestine where digestion continues until fatty materials pass into the lymphatic system while other food that has been digested enters the bloodstream.

All that stays in the small intestine is the undigested remains including water, harmless bacteria and parts that the body cannot digest, called roughage. These pass into the large intestine. About three-quarters of the water is absorbed here leaving the waste material to be passed out of the body. The waste matter is passed into the

The two kidneys and ureters drain urine into the bladder. The function of the urinary system is to produce urine, and thus wastes are filtered from the blood and transported from the body. The nephrons inside the kidneys filter between 170 and 200 litres of fluid in any 24 hour period. Nearly all of this fluid (about 99 per cent) is reabsorbed by the body.

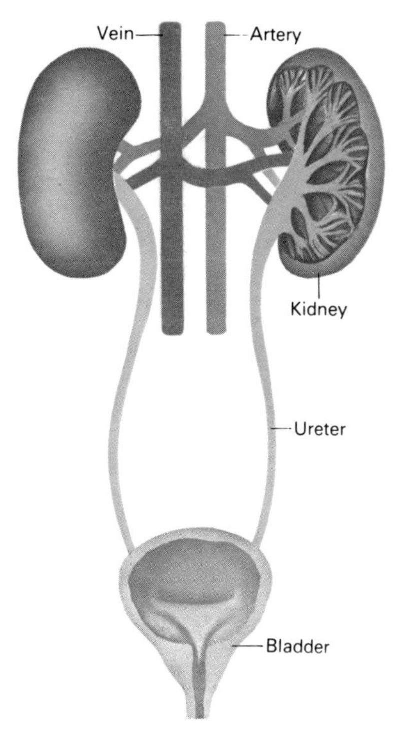

rectum at the end of the alimentary canal – the food channel that begins at the mouth and ends at the *anus* – and is passed out of the body as *faeces.*

The kidneys pass excess water and other liquid wastes out of the body and help to keep the proper chemical balance in the blood by eliminating excess salts. Other important *organs of excretion* are the lungs and skin. The lungs get rid of carbon dioxide when we breathe out, and the sweat glands in our skin excrete water and salt when we perspire.

GROWTH AND REPRODUCTION

Only living things can reproduce themselves. If you break a stone you get several small pieces but you will never have stones which grow to look like the original. Living things have young which grow to look like their parents. Humans reproduce themselves as babies which grow into adult humans.

Fertilisation and the embryo

In both animals and man, reproduction begins with a single cell, the same unit which makes up the whole body. Females make, within their bodies, special cells called *ova* (eggs). The male produces cells in his body called *sperm.* When a male sperm cell joins with an ovum inside the female's body, we say the ovum has been fertilised. The ovum grows and changes from one cell into billions of cells which form the earliest stages of a baby, called an *embryo.* The ovum does this by a process called *mitosis,* that is, the first cell splits into two and these go on splitting as the embryo grows.

Growth in the womb

This cell mass then divides into three sections: an outer layer, an inner layer called the *trophoblast,* and an inmost mass of cells from which the embryo grows. The growing embryo is attached to the wall of a special chamber inside its mother's body. This is called the *womb* or *uterus* and here it stays for the nine months it needs to develop into a baby ready to be born. This period is known as the gestation period. While the embryo is in the womb, it receives its food direct from its mother's body through the

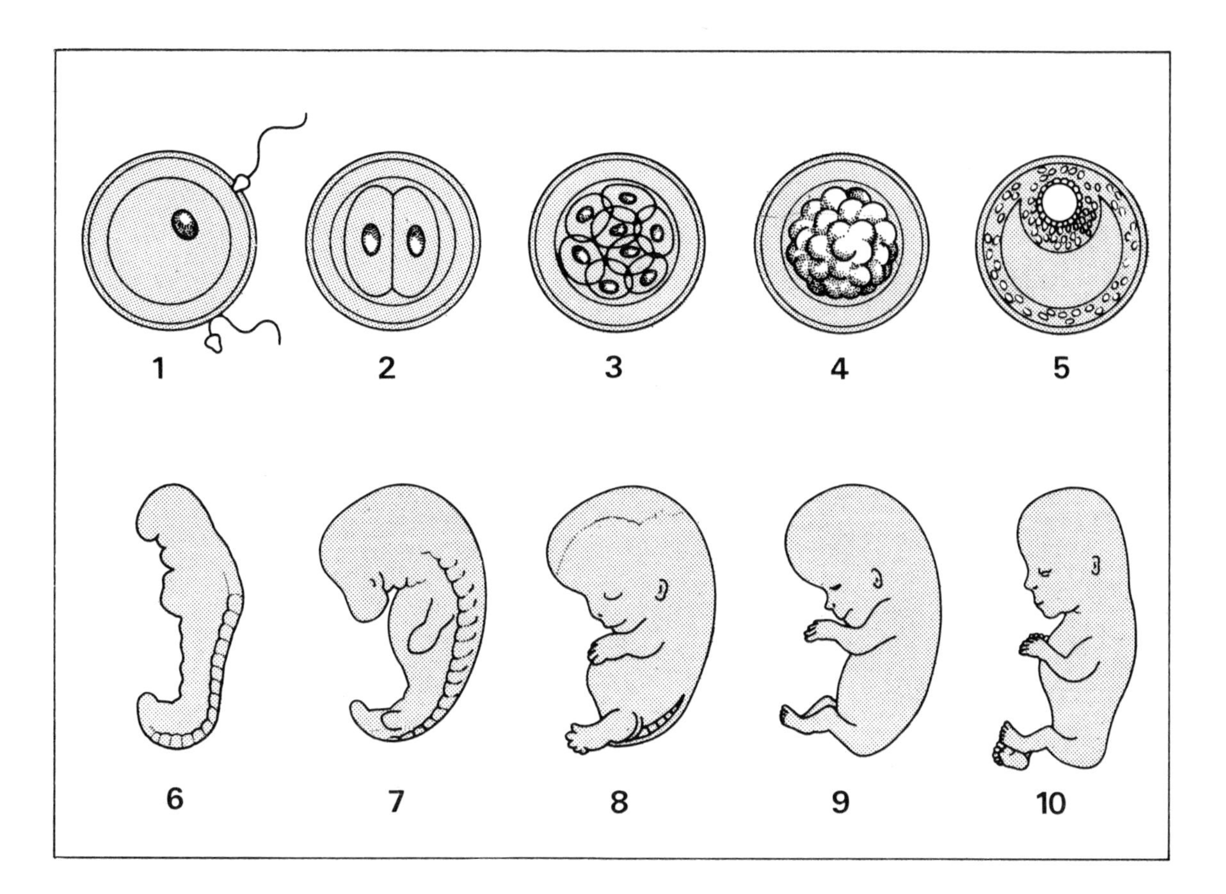

After the male sperm fertilises the female egg, the nucleus divides until an embryo begins to form (1-5). Its backbone starts to form at three and a half weeks (6), and at six weeks old, the tiny embryo has arms, legs and depressions where the eyes and ears will develop (7). The foetus is about 4 cm long at eight to ten weeks (8,9), its brain has developed and its limbs are distinguishable. At 12 weeks (10), the foetus is beginning to resemble a baby.

placenta. The embryo is joined to the placenta by a cord called the umbilical cord. When the baby is born, the doctor cuts this cord and the place where it was joined to the baby's body becomes the navel.

Wonderful changes take place in the embryo during its months in the mother's womb. From just three layers of cells in the embryo are formed all the body tissues and organs which make up the human body. This process is very complicated and takes place in stages. It is nearly complete when the embryo is about eight weeks old. During this time, the embryo has grown from less than the size of a pinhead to several centimetres long. In the next seven months until it is born, it will increase in length to nearly 50 centimetres and in weight from a fraction of a gram to about three kilograms. The baby is called an embryo usually until the early growth has taken place, then it is called a *foetus.* While the baby is growing inside its mother, we say the mother is pregnant.

When the baby is fully formed into a new human being, the muscles of the uterus push the baby out of its mother's body and the new child is born.

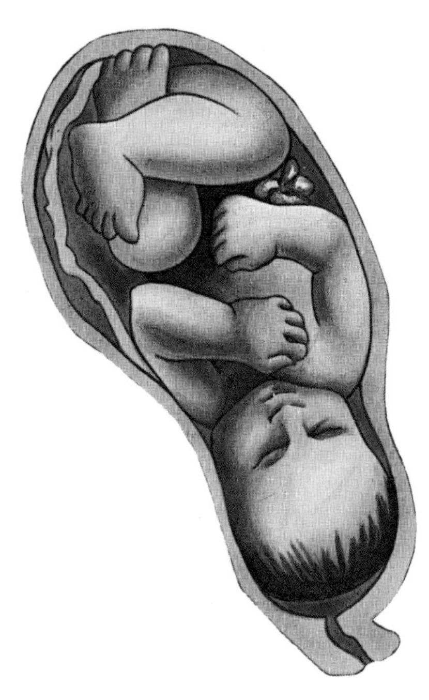

Twins

When two human beings or animals are born from the same mother at the same time they are called twins. In humans, twins are born about once in each eighty births. There are two types of twins. *Identical* twins look just like each other. They develop from a single egg which has been fertilised by a single male sperm. At a very early stage, this egg develops into two completely separate embryos.

When this separation is not quite complete and the two babies are born still joined together in some way, they are called Siamese twins because the first known babies born this way were born in Siam (now called Thailand). Identical twins are always either both boys or both girls. Non-identical or *fraternal* twins develop from two different eggs which are fertilised at around the same time, but these twins do not always look any more alike than ordinary brothers and sisters.

The fully developed baby in the womb obtains its nourishment from its mother by means of the placenta. Oxygen and food is passed continuously through the placenta to the baby and waste products are passed out into the mother's bloodstream. The baby is fully formed at 36 weeks and birth generally occurs 40 weeks after the mother's last menstrual period.

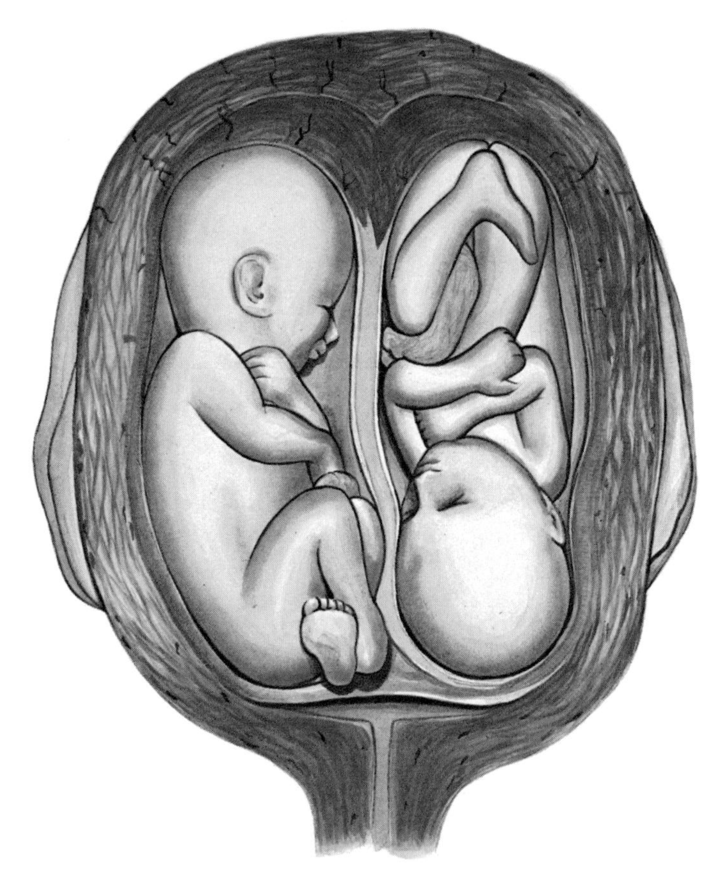

Right: Twins sometimes form and develop within the womb when an egg splits into two after fertilisation (identical twins) or when two separate eggs are fertilised by two different spermatazoa (fraternal twins). They share the same placenta but each foetus has its own amniotic sac.

These Swedish girls have all inherited their parents' blue eyes and fair hair. Certain physical features are passed on genetically from parents to children. The nucleus of every body cell contains about 100, 000 genes and 46 chromosomes, and thus when an egg is fertilised by a sperm to produce a human foetus, the resultant baby inherits certain genes from both of its parents.

Heredity

When an aunt or uncle tells you 'you have your father's nose' or 'your mother's blue eyes' they mean that you look a little like your parents in these ways. This is called *heredity* because you have *inherited* or received some things from your parents. Children get some of their outward appearance from their mothers and some from their fathers. Things such as height are inherited; usually two short parents will have children who do not grow very tall. The study of the way in which characteristics are handed down from parents to their children is called *genetics.* Inheritance is controlled by bodies called *chromosomes,* which are present in the male and female cells and carry what are called *genes.* The genes a person receives from his parents decide his physical characteristics and such conditions as haemophilia.

Growth

Growth is the gradual increase in size and weight of all living things. After the baby is born, it goes on growing, usually until about the age of 18 years, when the person is adult. All the parts of the body do not grow at the same rate. The arms and legs usually grow much more quickly than the body and the head. Plants and some simple animals can also grow back a part that is lost. This is known as *regeneration,* but a person cannot regrow a lost finger in the way a lizard can a new tail.

Growth is affected by chemical substances in the body called *hormones.* These hormones are made in the body in

This illustration shows the position of the different glands in the human body. There are two types of glands: ducted exocrine glands and ductless endocrine glands (cross-sections of each are shown below). The endocrine glands produce the hormones that control many of the body's functions and growth.

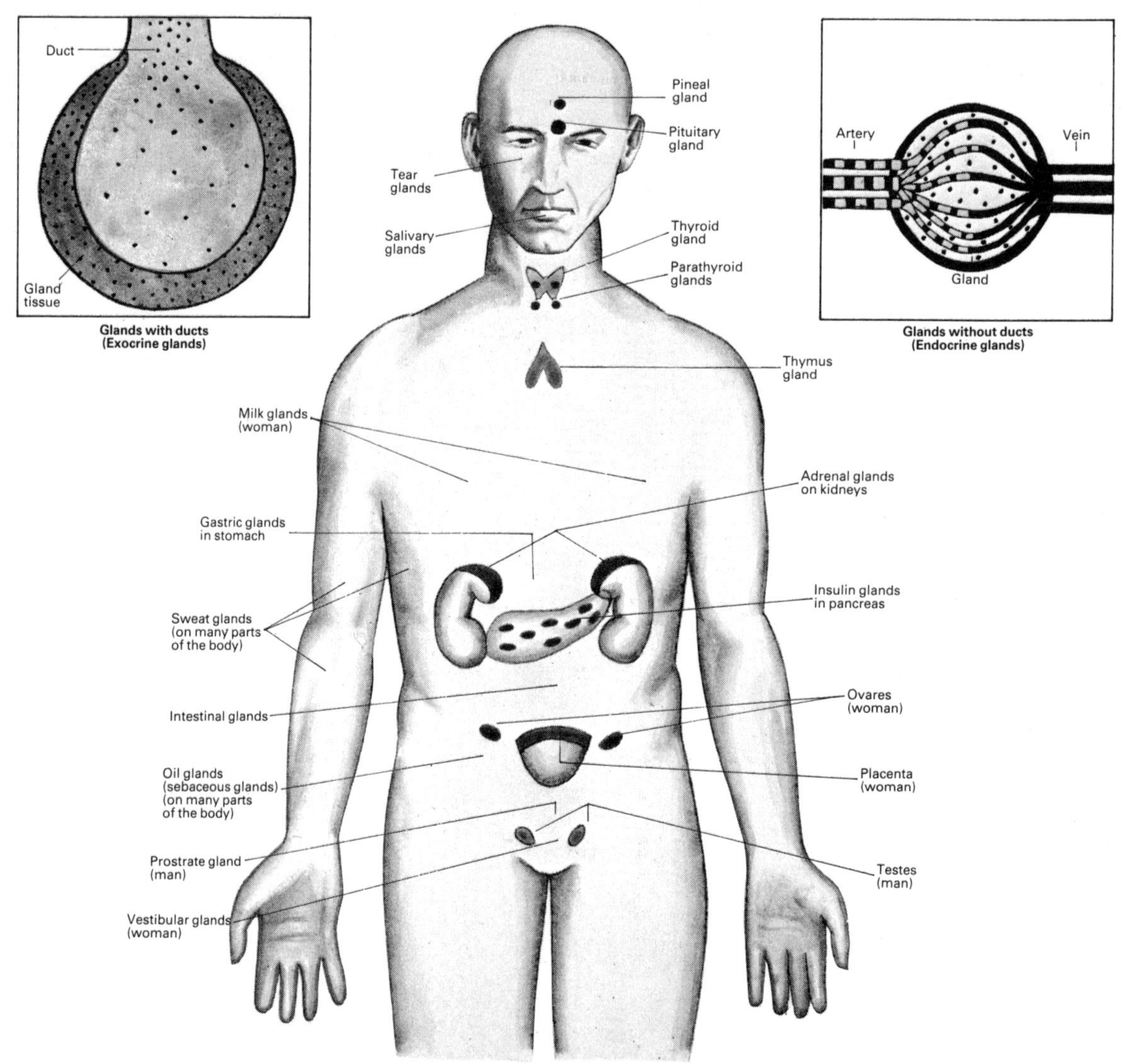

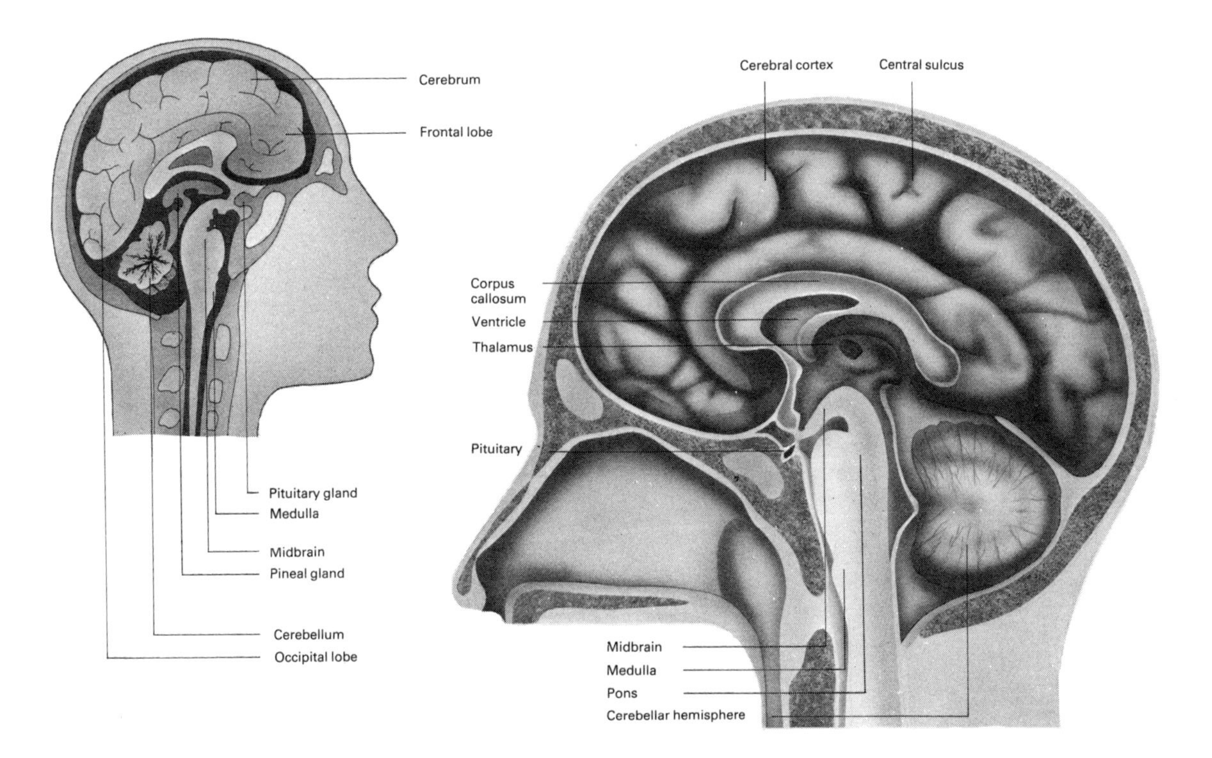

Man's brain is encased within his protective bony skull. The brain, which weighs from 1250-1380 gm, is made up of 30,000,000,000 cells and continues to grow until the individual is 20 years old. A mass of nerve fibres connect the cells and transmit signals from the body to the brain and from one part of the brain to another, information being transmitted by means of minute electrical impulses.

organs called *glands.* The hormones are the body's messengers. They are slower-acting than nerve signals but they can act for a longer time and over a wider area. A most important gland is the *pituitary gland,* which produces the hormone that controls growth.

Other important glands are the *thyroid gland,* which controls the body's burning of food to give you energy; and the *adrenal glands.* These last glands help the body to get ready for action when we are faced with sudden danger, or when we are excited about something. The adrenal glands send out hormones which increase heartbeat and breathing, and send sugar into the bloodstream to provide energy in case we have to fight or run away from whatever the danger is. That is why you find yourself breathing very fast and can feel your heart pounding when you are frightened of something.

The brain and nerves

The brain controls the activities of the body and forms the chief part of what is called the *nervous system.* From the brain a column called the spinal cord runs down inside the bones or vertebrae that form the spine. Thirty-one pairs of nerves run off the spinal cord to different parts of the

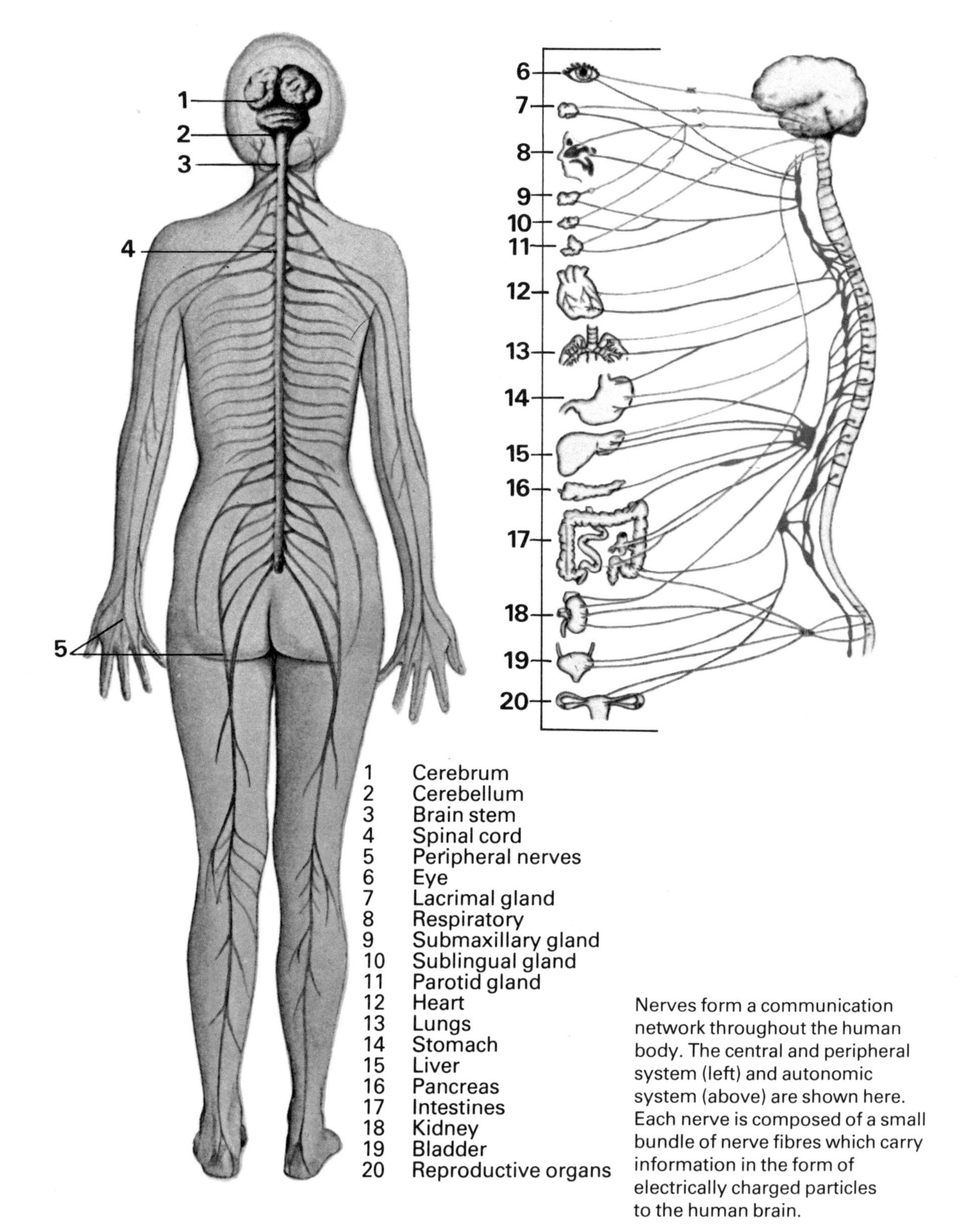

Nerves form a communication network throughout the human body. The central and peripheral system (left) and autonomic system (above) are shown here. Each nerve is composed of a small bundle of nerve fibres which carry information in the form of electrically charged particles to the human brain.

body. Nerves are of two types: the *sensory* nerves, which carry messages from the senses, and *motor* nerves, which carry the impulses, or orders, from the central system to the muscles.

What the brain does

The brain receives information through the senses of sight, hearing, smell, taste and touch. The organs of these senses- the eyes, ears, nose, tongue and palate -and skin all send their messages about what we experience along the nerves to the brain. With the brain we are able to decide conscious reactions to our experiences. When we see a car approaching, we do not step into the street, but decide to wait till it passes. Some sudden or dangerous experiences provoke an immediate unconscious reaction; for example, if we touch a hot object and jerk our hands away, this is called a 'reflex' action, because the message from the sensory or 'receptor' nerve is passed directly to the motor nerve.

When we know or are *aware* of our brain acting to control our body or our thoughts and feelings we are awake or *conscious* and able to think, decide and act consciously. But many activities of the body are not controlled consciously. In everyday life these are especially the breathing, the circulation of the blood and the processes of digestion of food. The messages received from both inside and outside the body must be co-ordinated and this occurs in the *central nervous system,* which consists of the brain and spinal cord, while the

The electroencephalogram (1) can measure brain wave patterns and these are represented on a screen. The waves on the screen (2) correspond to the different states of being awake and relaxed, awake and concentrating, and ordinary sleep.

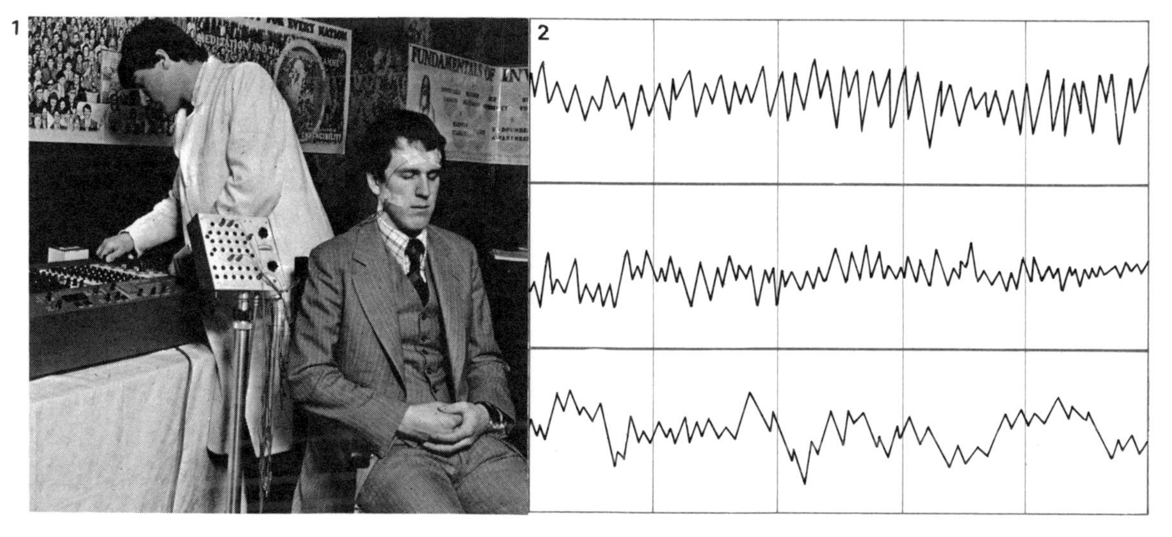

nerves that control our unconscious processes are called the *autonomic nervous system.*

The brain also makes it possible for us to retain memories of what we experience and to recall these memories. With the brain, man is able to learn and to use his experience.

The simplest animals have no brain, but some have a very simple nervous system which controls their bodies. All vertebrates or backboned animals have brains which are somewhat alike. Mammals and birds have well developed brains, and man's is the most highly developed of all.

The brains of all creatures are made up of nerve cells and fibres. The human nervous system has many millions of nerve cells and most of these are in the brain. Each one of these cells connects to many others so that a single nerve signal may be passed onto different parts of the brain.

These diagrams show the different brain structures of a mammal and a bird. Whereas the mammal's brain has a well-developed cerebral cortex, the bird has an 'instinct' brain. Thus the corpus stratium, which coordinates instinctive patterns of behaviour, is highly developed. The mammal is a much more intelligent animal than the bird and this is reflected in its more complex brain structure.

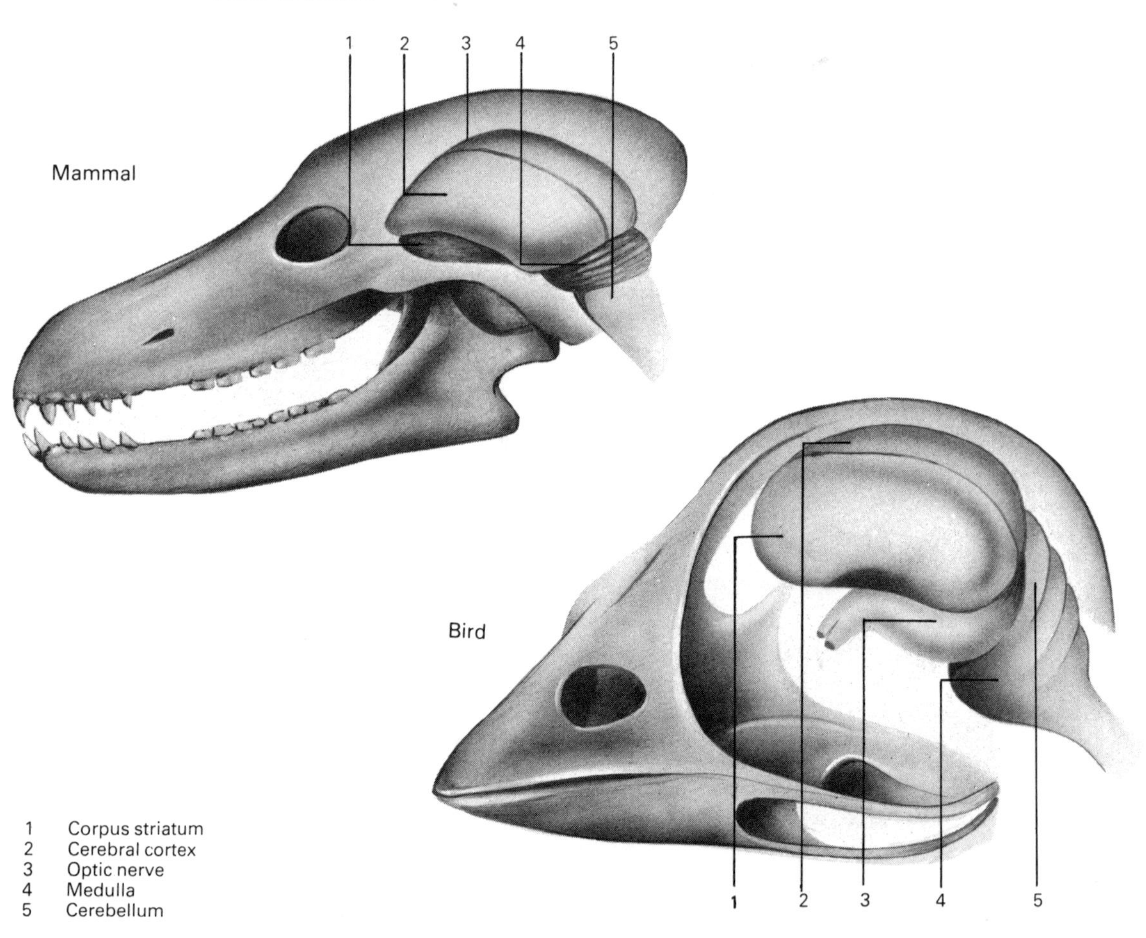

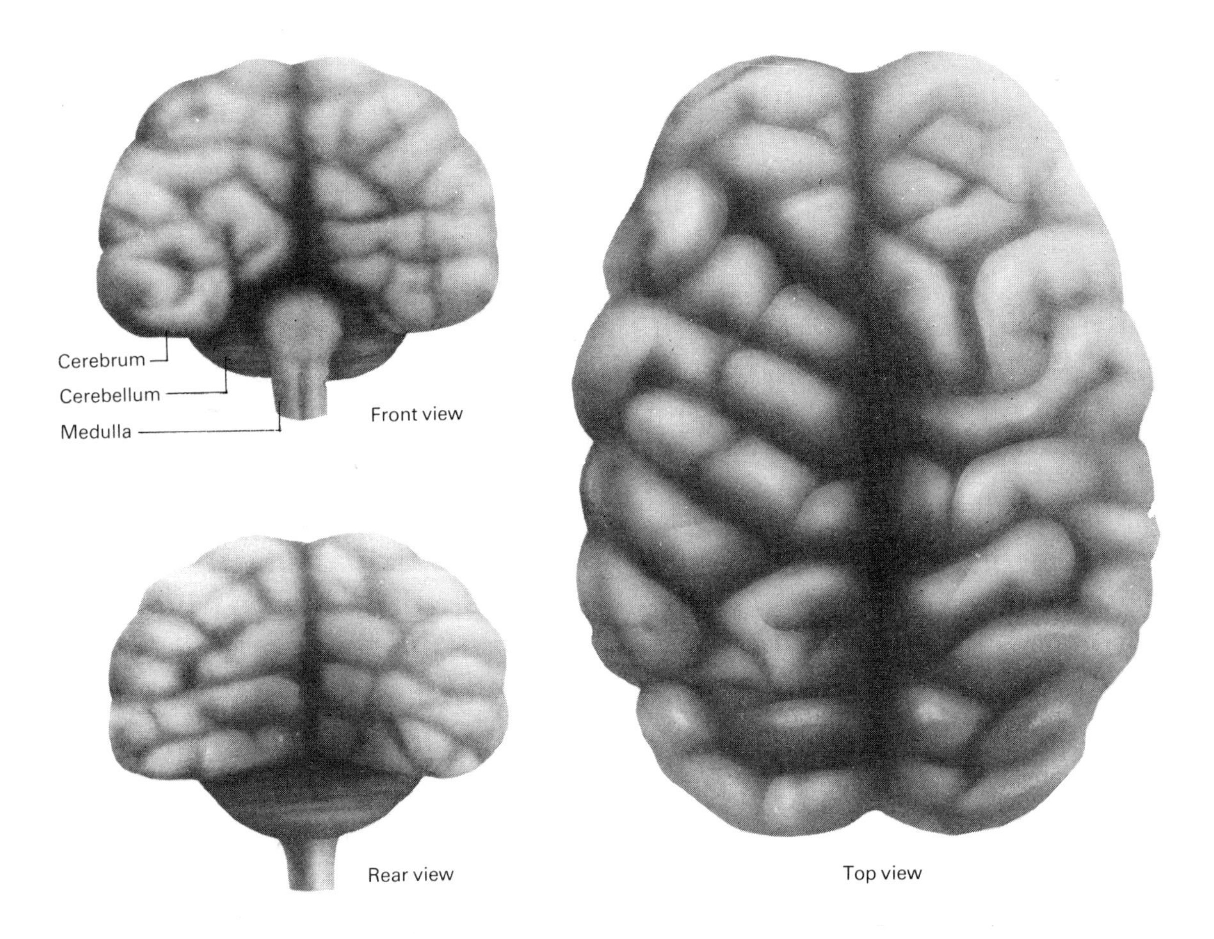

The structure of the brain

The human brain consists of a 'stalk' which is the top part of the spinal cord, with three large swellings attached. At both sides of the top and at the front are two areas called cerebral hemispheres. These areas are the most highly developed parts of the brain and they control our conscious responses to the messages of our senses. The outside of these hemispheres is very deeply wrinkled and has an outer covering of *grey matter,* called the *cerebral cortex.* This is made up of a mass of nerve cells all linked with one another. If a map is drawn showing this cerebral cortex, we can see that different parts control different activities. The parts at the back, for example, known as the *occipital lobes,* receive messages from our eyes. The cerebral cortex is the centre of all these processes which together are described as intelligence.

Each half of the brain controls one side of the body, but the nerve signals cross over to the opposite half or

When the brain is viewed from above, the two halves, which are joined by the corpus callosum, can be seen clearly. The right half of the brain controls the left side of the body and the left half the right side. But although they are mirror images of each other, they each have different functions. Thus speech is controlled by one half of the brain and intuitive activities by the other.

hemisphere. This means that the left half of the brain controls the working of the right side of the body and the right half of the brain controls the working of the body's left half. Usually one half of the brain has more influence than the other. In most people, this is the left side of the brain, which is why most people use their right hands more easily than the left.

Behind and below the two hemispheres is the third swelling, called the *cerebellum.* This part of the brain helps you keep your balance and coordinate your movements. Another part called the *hypothalamus* is important in keeping your body at the correct temperature and controlling your appetite. The *medulla oblongata,* which is the thickened upper part of the spinal cord, controls the unconscious bodily processes like digestion.

The visual nerve fibres cross over and although the retina of each eye has a total image of the object, the left half of the brain sees only the right part of the field of vision and the right half only the left part. The nerve cells in the retina send information on colour and intensity of light to the brain.

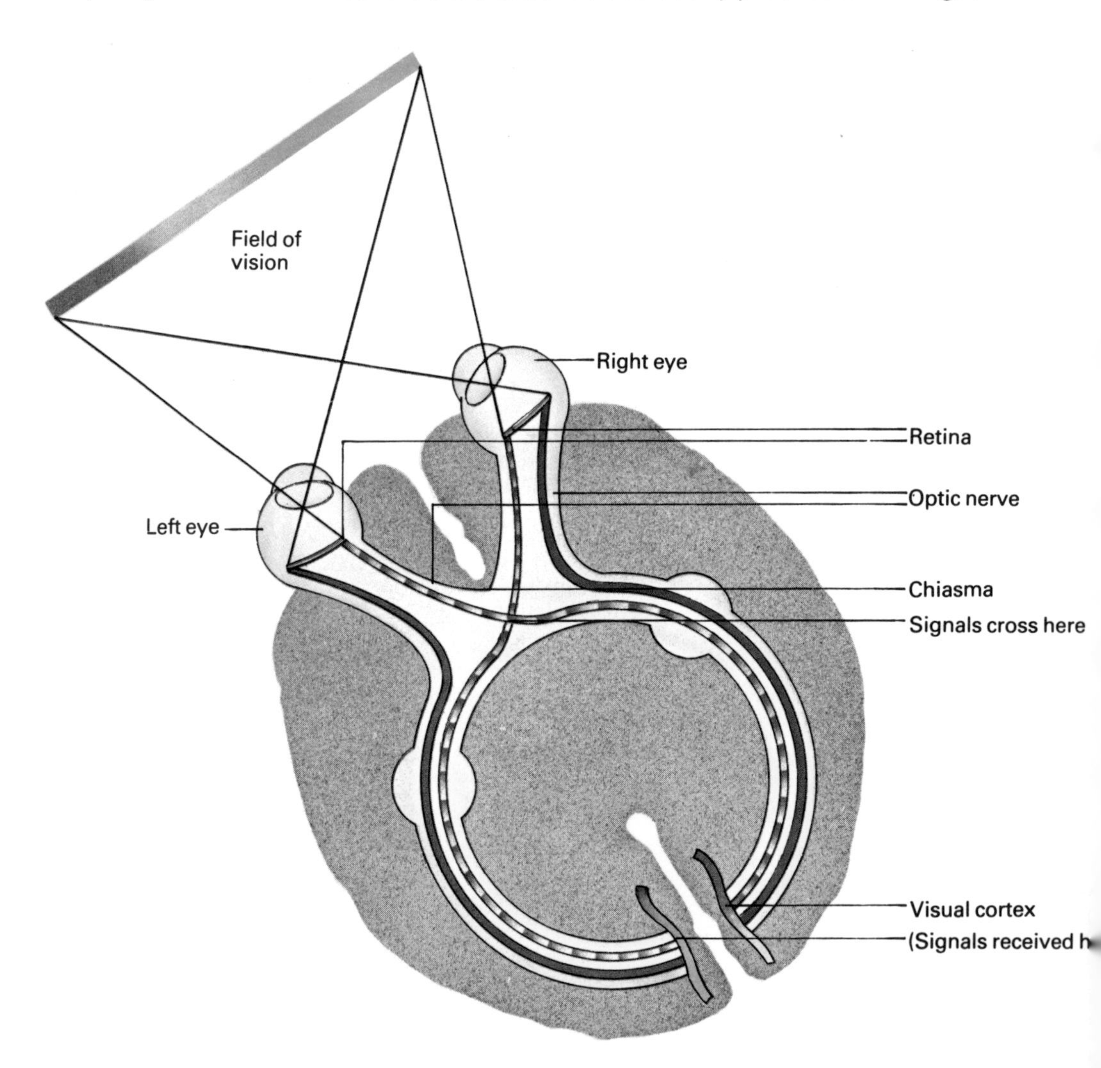

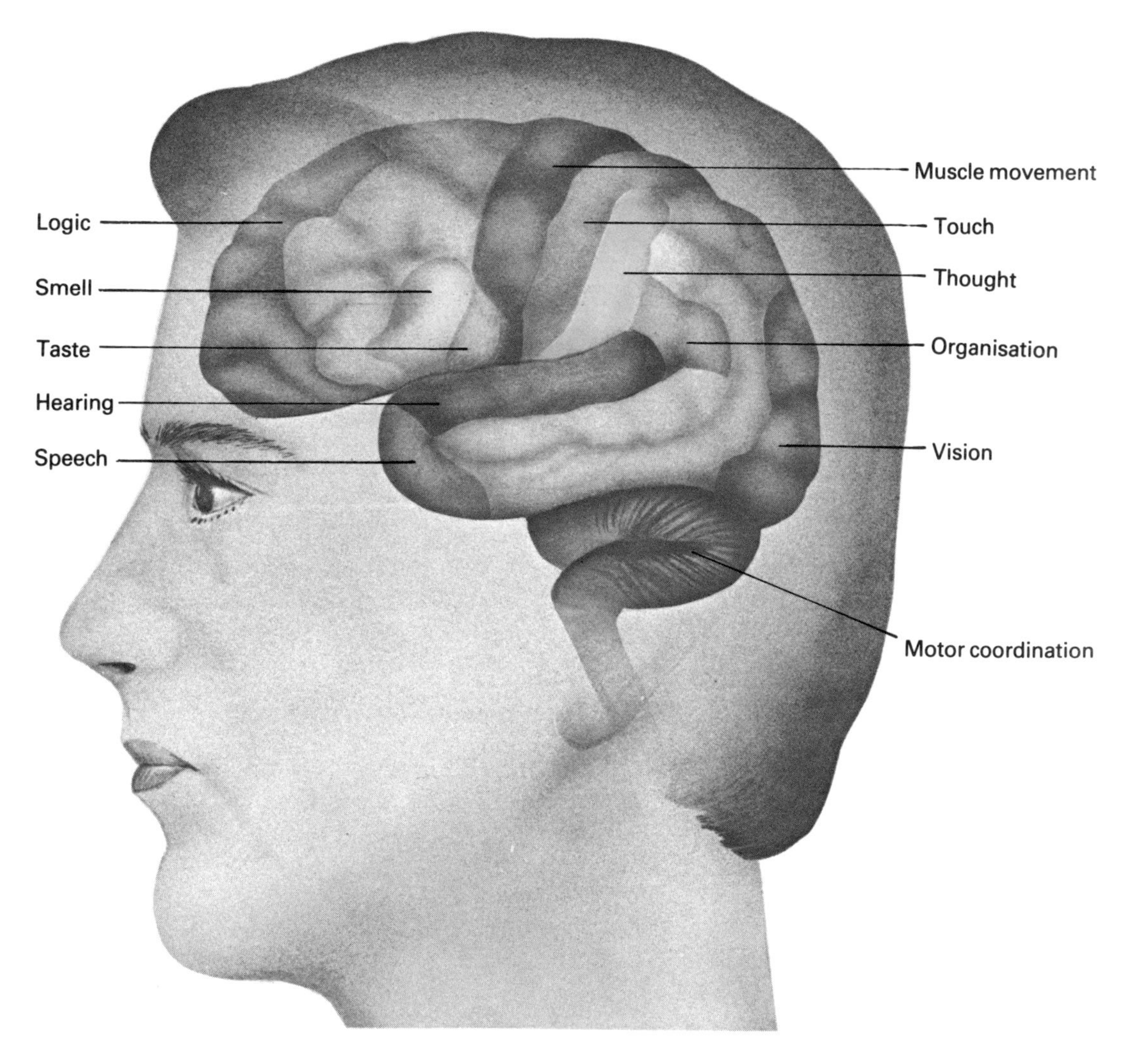

Brain damage

Because nerve cells can never be replaced after their death, the brain is easily damaged and the effect depends on which part is damaged. Any accident such as a heart attack or choking which stops oxygen reaching the brain for more than a few minutes will damage the brain. If the left side of the brain is harmed, a right-handed person may be very badly affected, while damage to the other side may have less effect. When someone is left-handed, it is because the right half of the brain is dominant. Left-handedness is found in less than ten per cent of people. It is very common in twins, but usually only one twin of a pair is left-handed.

Different areas of the brain are responsible for such functions as thought, vision, muscular movement, coordination, logic, smell, taste and personality. The right hemisphere of the brain is involved in artistic and imaginative activities whereas the left side is mainly concerned with more logical, analytical pursuits: critical thinking, numbers, language.

INTELLIGENCE, LEARNING AND SLEEP

The word *intelligence* describes a person's overall mental ability to carry out various tasks and solve problems. It includes the ability to learn new things and remember them, to solve problems and use past experiences to help deal with new situations.

Intelligence tests

Tests, called intelligence tests, have been invented to compare the intelligence of different people. The result of an intelligence test is given as an *intelligence quotient* or IQ. By comparing the scores of many different people of the same age it is found that most people have an IQ of between 90 and 110, and this is taken as average intelligence. Only about 5 per cent of people tested have an IQ of less than 70 or more than 130.

The instincts of hunger and feeding are common to all animals, including Man. But although it is instinctive to this baby to eat, he has to learn how to eat in a socially acceptable way, and mastering the art of using such feeding implements as spoons and forks is a form of learned behaviour.

Writing is a form of learned behaviour, which is taught to children at school. Young children operate on a concrete operational level, which means that they can use rules and concepts in their concrete form but have still to master their abstract use. Having learned to write and read, the children never forget, and the skill is stored in their memory for the rest of their lives.

Learning and memory

There are two main causes for the actions that living creatures perform. These are *instincts,* which are born in us; and *learned behaviour* which we are taught or teach ourselves. A bird does not have to learn how to build a nest; it does this instinctively. But no human baby is born with a knowledge of arithmetic; this has to be learned. Instincts are not so important to people as they are to animals and so nearly all our behaviour is learned.

When you repeat something you have learned, like spelling a difficult word, you know how to do it again because you have stored the information in your memory. One way of remembering something is to repeat it many times and it is also easier to remember something if you find it interesting. Wanting to learn something is also important, this is called *motivation.* If you want to get good marks in a test, that is your motivation for learning the subject.

One type of learning is known as *conditioning.* It was discovered in 1902 by a Russian, Ivan Pavlov, who found that if he rang a bell at the same time he gave a dog some food, the dog produced saliva in its mouth, as we do when food 'makes our mouth water'. Soon, ringing the bell was enough to make the dog's mouth water even though no food was put out. Producing saliva when food is offered did not have to be learned, it was instinctive, or natural to the dog. However, producing saliva when a bell was rung was something the dog learned.

However we learn something, the experience must be stored. Just how the brain does this, we do not know. But we do know that there are two kinds of memory; a kind of temporary store where we keep unimportant memories such as a telephone number; and a permanent memory store where we remember more important things. When some accident, such as a blow on the head, damages the working of this memory storage, we may not be able to remember some things for a time. This 'loss of memory' is called *amnesia.*

Sleep

About a third of our whole life is spent asleep. This seems to be when the body repairs itself. Although the activities in our bodies slow down during sleep, our brain is still active although we do not think consciously. Scientists have been able to study the electrical activity in the brain through a machine called an *electroencephalogram,* called EEG for short. This shows that there are two different kinds of sleep and we usually pass from one kind to the other several times in one night.

The deepest kind of sleep is non-dreaming sleep. The second type is much lighter, and dreaming takes place during this type of sleep. Scientists have found that, during dreams, the person's eyes are constantly moving although the eyelids are closed, as we 'watch' the action in the dream.

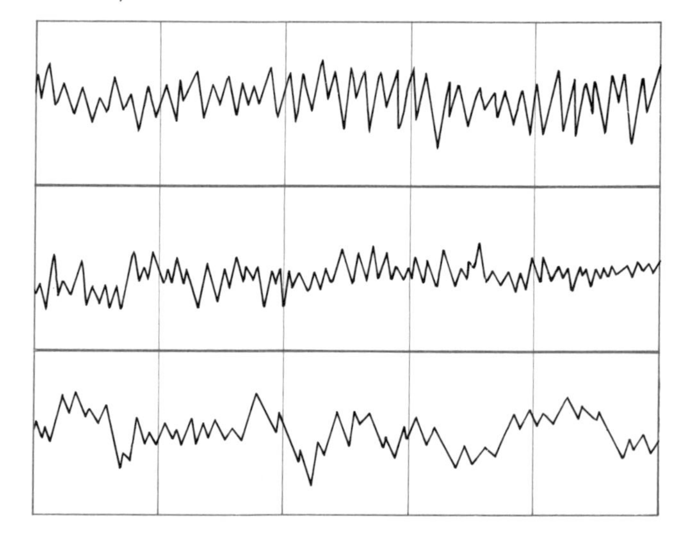

An EEG machine can 'see' the way in which we sleep and a pen traces the information on to a moving paper belt. The regular alpha rhythm waves (waves with a frequency of 10 cycles per second) change as a person falls asleep to irregular waves and, later, short waves interrupted by bursts of faster waves, which are called 'spindles'.

THE SENSES

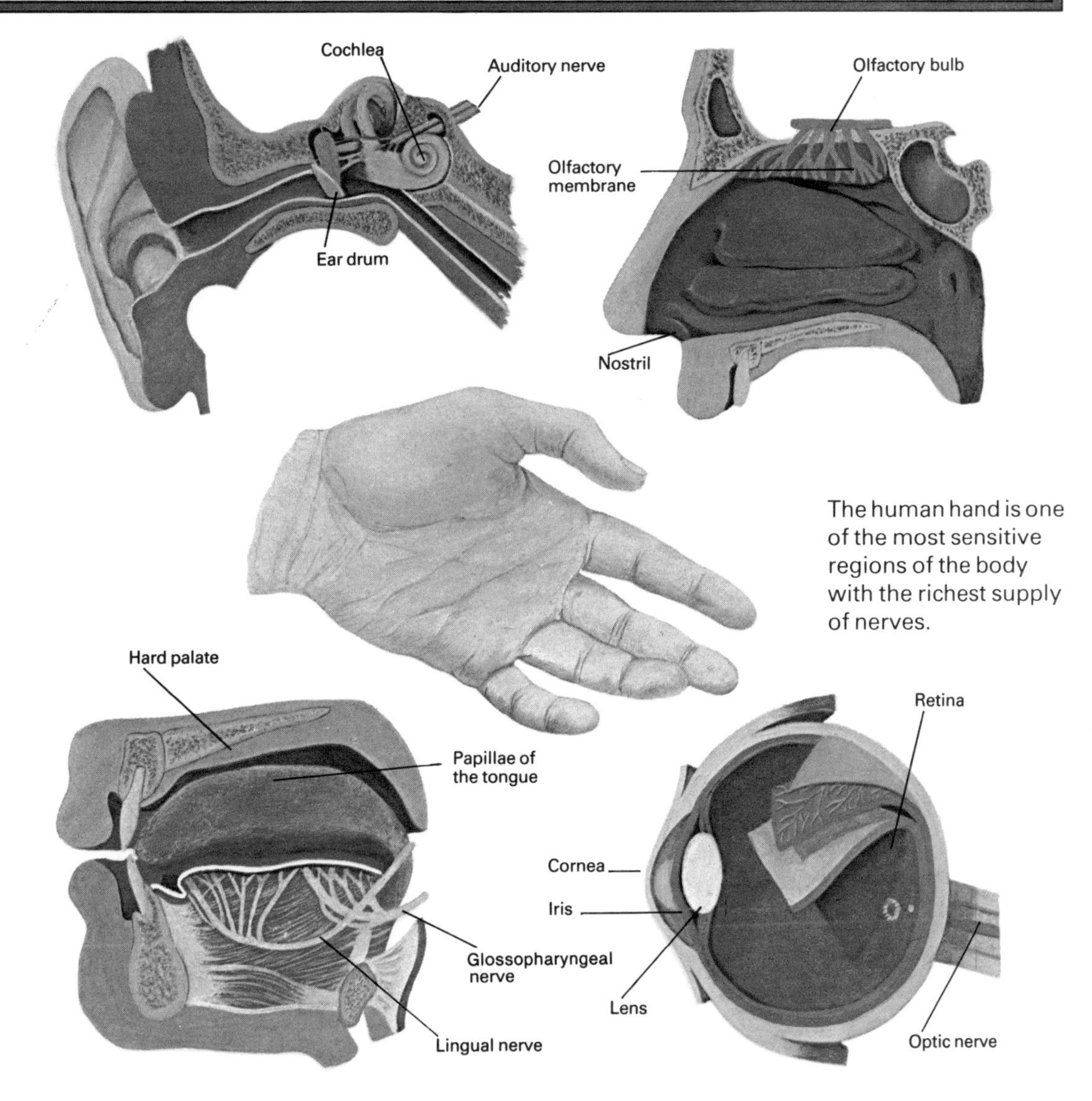

The human hand is one of the most sensitive regions of the body with the richest supply of nerves.

Our senses let us learn about our surroundings and about ourselves. The best known senses are sight, hearing, taste, smell and touch. These were called the 'five senses' for many years, until scientists added the senses of heat, cold, pain and pressure. These are external senses which tell us about our bodily needs and give us information about our body. They include also the sensing of hunger, thirst, tiredness, pain, balance and our body's position and movements.

These cut-away views of the ear, nose, mouth and eye show their internal structure. We use the five senses — sight, smell, touch, hearing and taste — to interpret the world around us, and receptors in these organs send nerve impulses back to the brain for interpretation.

Stimuli and receptors

The changes in the world around us which are detected by our senses are called *stimuli.* For each sense, we have sense organs which respond to stimuli by sending nerve signals along paths of sensory nerves to the brain.

When signals from the senses reach the part of the brain called the *cerebral cortex,* we become aware of them. Depending on what caused the signals, we can sort them out as smell, taste, sound, touch or whatever.

Our most sensitive and accurate sense is sight, and the *receptors* or nerve endings are the *rods* and *cones* in the *retinas* of our eyes. Light is the stimulus which makes these receptors produce the nerve signals that we interpret as seeing.

The organs of hearing are the ears. They are stimulated by the vibrations in the air which we call sound. The receptors which produce the nerve signals are delicate organs in the inner ear.

The eye has a slight bulge at the front and a stalk containing the optic nerve behind. Light enters the eye through the transparent cornea and is focused by the lens on to the retina, which contains millions of light-sensitive cells called rods and cones. The pupil in the centre of the coloured iris controls the amount of light entering the eye and can enlarge or dilate according to the intensity of the light. Thus in dim light, the pupil expands, whereas in bright light it gets smaller so that less light enters the eye.

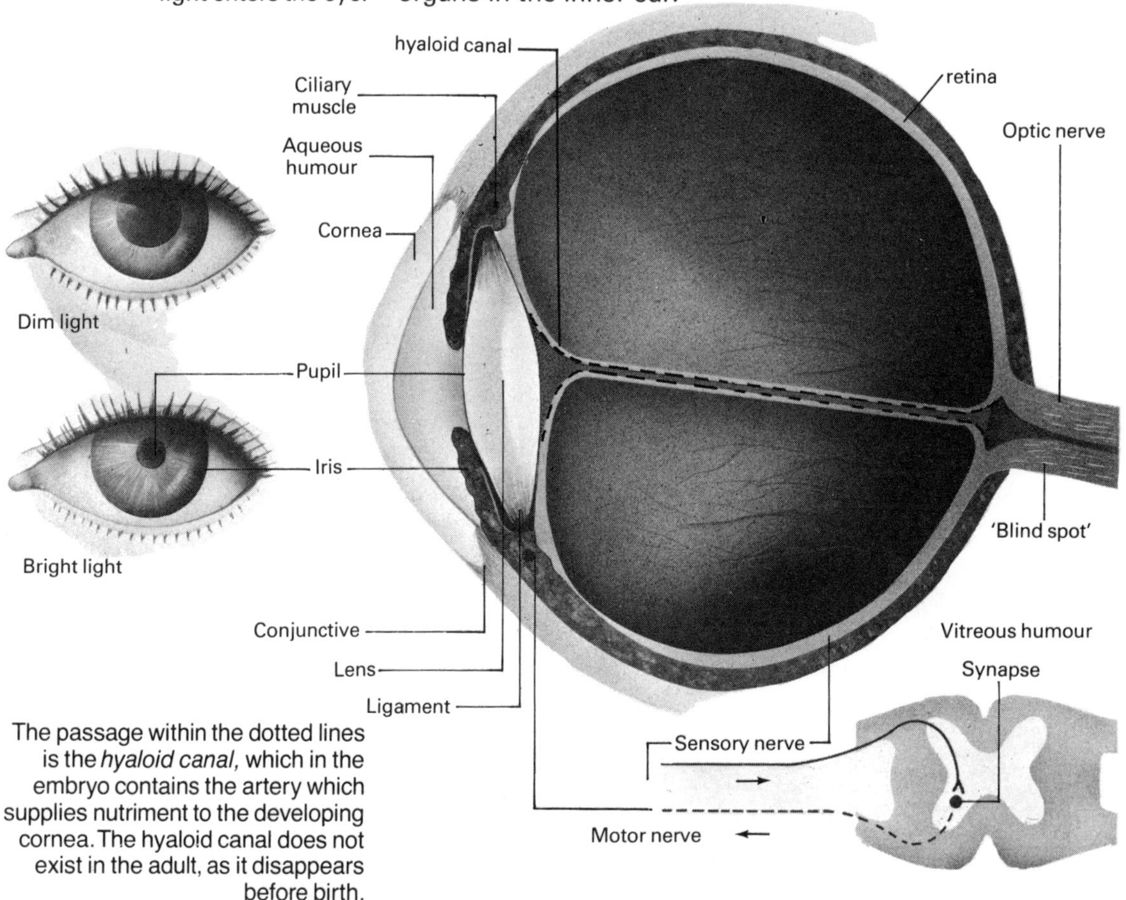

The passage within the dotted lines is the *hyaloid canal,* which in the embryo contains the artery which supplies nutriment to the developing cornea. The hyaloid canal does not exist in the adult, as it disappears before birth.

Taste and smell

Taste and smell are called chemical senses because they respond to chemical substances in what we breathe, eat and drink. For many animals, like the dog, smell is one of the most important of all senses. They use it to find food and mates and to detect danger. One type of male silkworm moth is known to be able to detect the scent of a female over a kilometre away. In man, the sense of smell is not so well developed, although it is a great deal more sensitive than our sense of taste. Smells are detected by special receptor cells high up in the part of the nose called the nasal passages. These cells are a special type of nerve ending. They are packed closely together in about 10 square centimetres of the nasal lining on each side. They send messages along the *olfactory nerve* to the part of the brain called the olfactory, or smelling lobe.

Most of what we call the flavour of food is really its smell. The receptors for taste are tiny *taste buds* in the mouth. There are four main kinds of taste buds, for salty, sweet, sour and bitter tastes. The tip of the tongue is the most sensitive. An adult has about 9,000 taste buds but a baby has many more. Not all of these are on the tongue. They are also found on the walls of the mouth and on the throat. Each taste bud has a nerve ending set into a tiny pit.

Taste and smell are separate chemical senses, which are closely linked. Thus a person complains that he is unable to 'taste' food when suffering from a head cold. There are about 10, 000 taste buds (1) on the surface of the tongue, and when we eat, the food chemicals react with the sensory hairs. The sensitive scent cells in the nose detect any odours (3) in the air. Wine tasters (2) have a highly developed, acute sense of both smell and taste and thus can identify hundreds of wines.

1

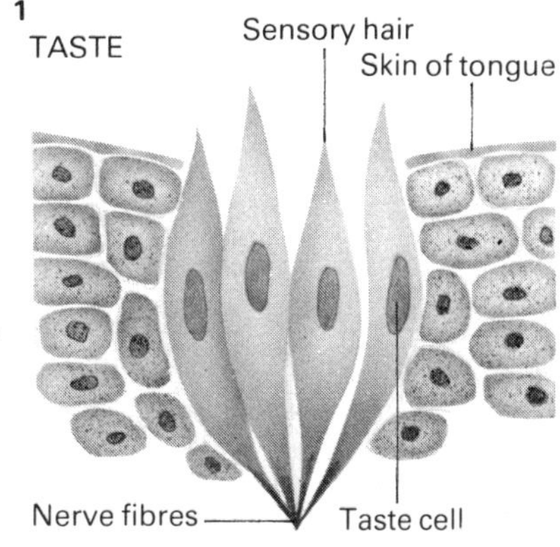

2

3

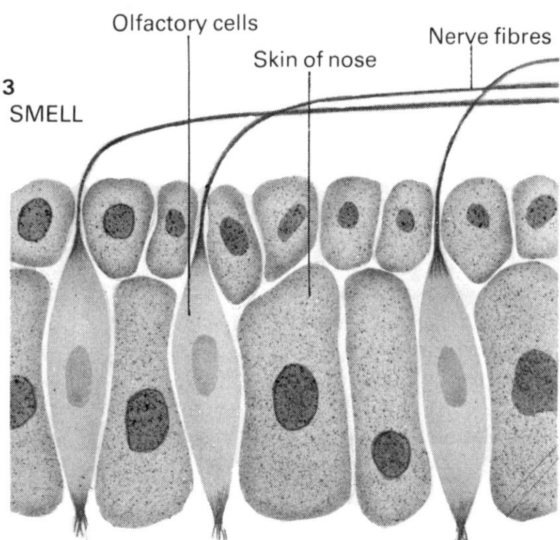

Touch

We recognise the sensations of touch, pain, cold, pressure and heat by means of the millions of minute nerve fibres in our skin, which send back messages to the brain. This cross-section through the skin shows the various nerve fibres. These sensory perceptors are sensitive to different things. For example, the Ruffini corpuscles react to heat, whereas the Kraus end bulbs, most prevalent in the tongue and the eye, can sense cold.

Touch actually includes several senses and all of these respond to different stimuli applied to the skin.

The sense of touch is very important because it gives information about what is happening to the outside of the body and warns of possible dangers. If we touch something sharp, our nerves send a 'danger' signal and we pull away even before we have time to think about it. This protective action is called a reflex action. The sense of touch is not as sensitive as our other senses and we usually need strong stimuli before we react.

The different forms of touch sensation, pressure, pain, heat and so on, are recorded by nerve endings in our skin. Nerve signals from these touch receptors are sent to part of the brain called the *parietal lobe.* Here there is a part of

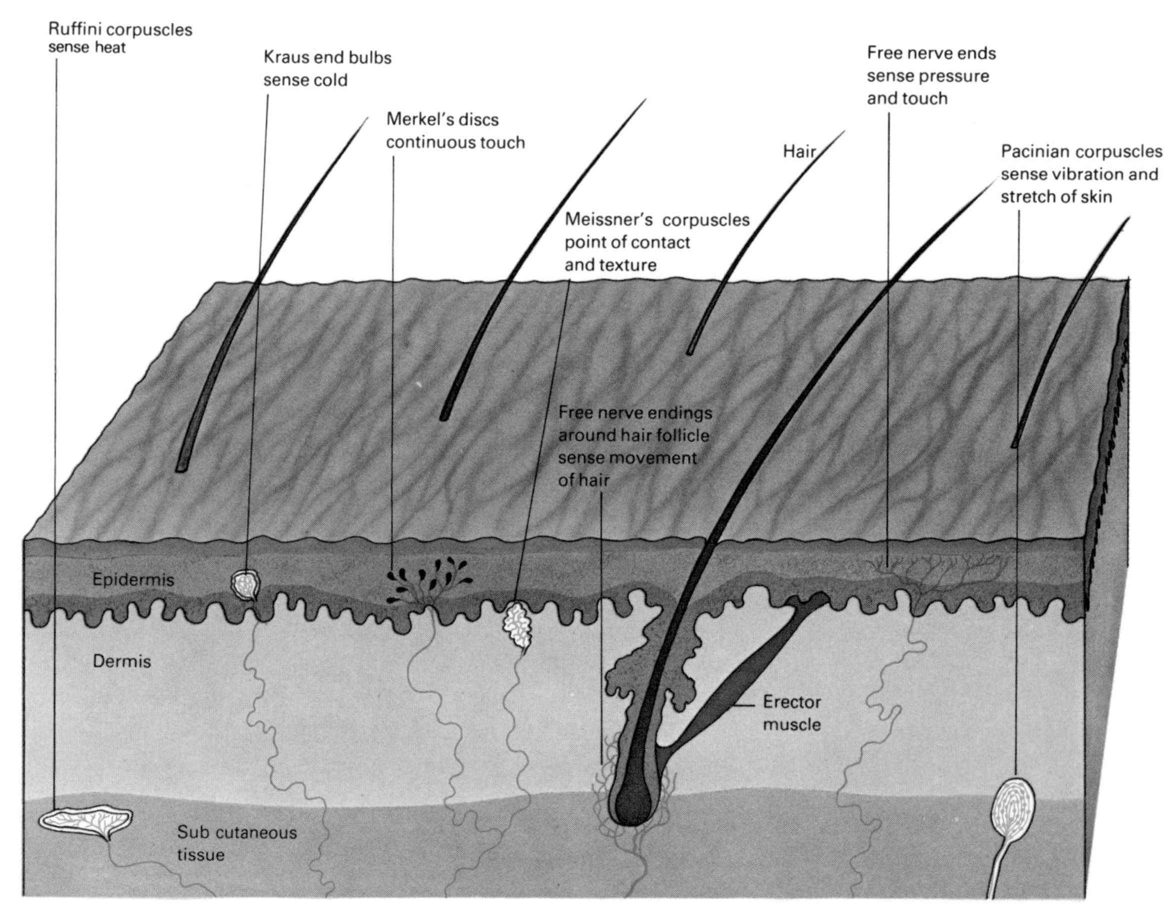

This blind woman is reading braille, a system of reading based on a special pattern of raised dots. She uses both hands to read it, and the extra sensitive nerve fibres in her fingers quickly touch the dots and transmit the information to her brain for decoding.

the brain's cerebral cortex which corresponds to each area of the body, so the brain knows where each signal has come from. The most sensitive parts of the body, such as the fingers and lips, send the strongest signals to the brain.

Internal senses

Our internal senses which tell us about our body are also important. For example, we know the positions of our arms and legs without looking down to see whether they are folded or crossed. Our sense of balance is also internal, and there are some pain receptors inside the body, although not as many as on the skin.

Hunger and thirst

These senses tell us when we need to eat or drink. Scientists now believe that hunger and thirst are separate senses, although their centres are close together in the brain. Two things seem to trigger our sense of thirst; if the amount of fluid surrounding our body cells drops, or if

Rays of light enter the eye and pass through the cornea before being bent by the lens and focused on the retina. An image of the observed pencil is thus formed on the retina. The cones in the retina distinguish its colour and information is carried to the brain by the optic nerve. The brain perceives distance and depth.

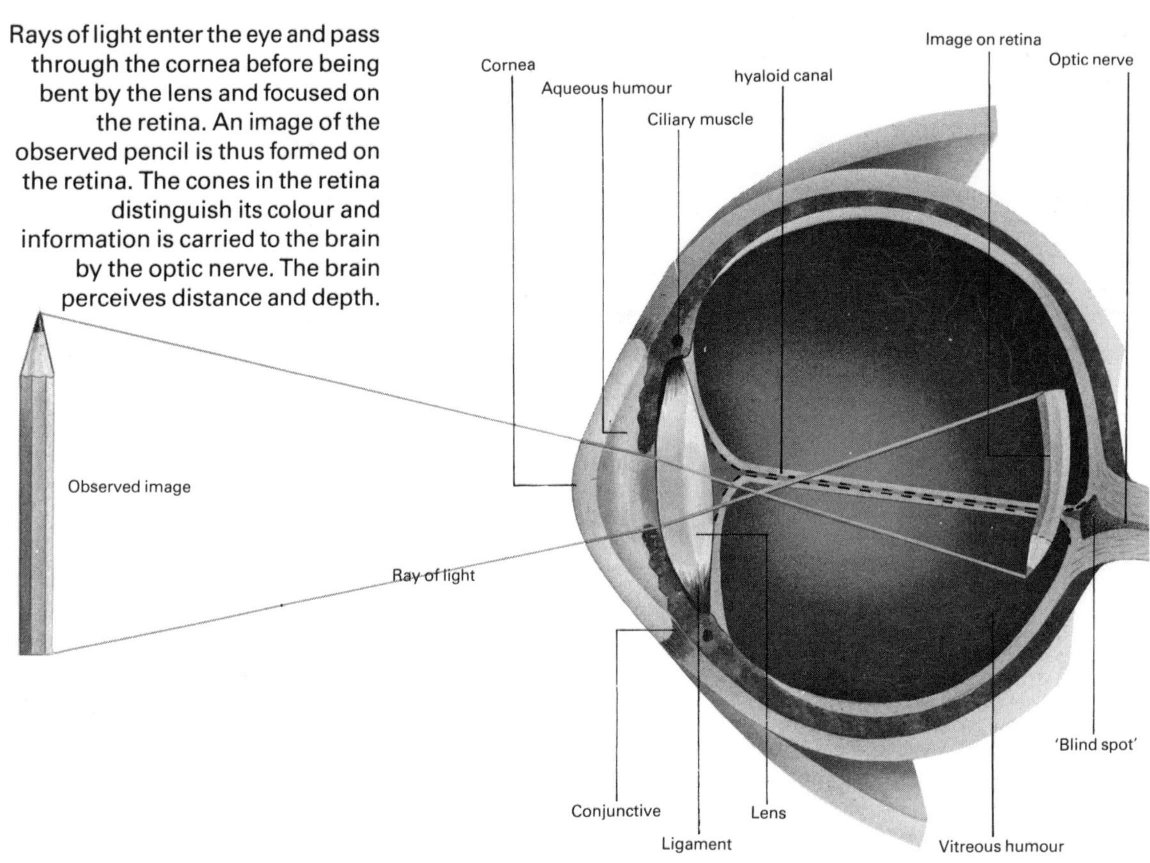

there is an increase in the blood's saltiness. This is why drinking sea water makes us more thirsty instead of less.

An area of the brain called the *lateral hypothalamus* seems to be important to both hunger and thirst, because if this area of the brain is damaged, the person refuses both food and water. If the nearby part of the brain, the *ventromedical hypothalamus,* is damaged, a person will eat as much food as is available, without seeming to realise when he has had enough to eat. Normally, a special hormone called a *satiety hormone* signals when the body is satiated or satisfied. It is released into the bloodstream by the stomach.

Short-sightedness occurs when the eyeball is too long and the lens too weak. This condition can be easily corrected by a convex lens. Long-sightedness can be corrected by using a concave lens to focus the light rays on to the eye's retina.

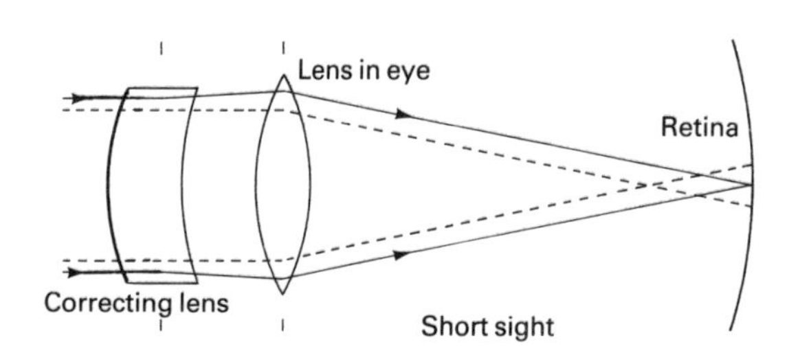

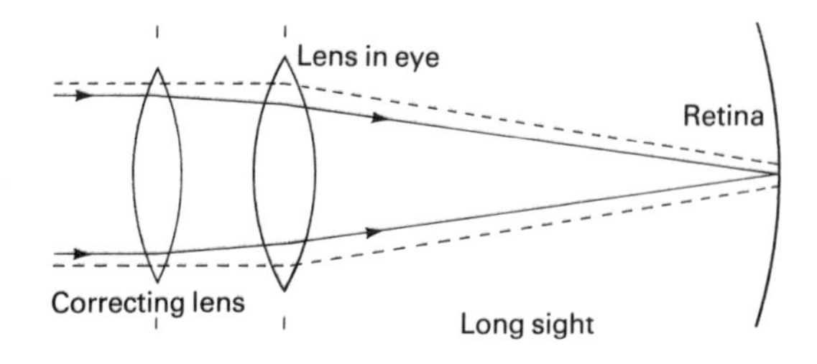

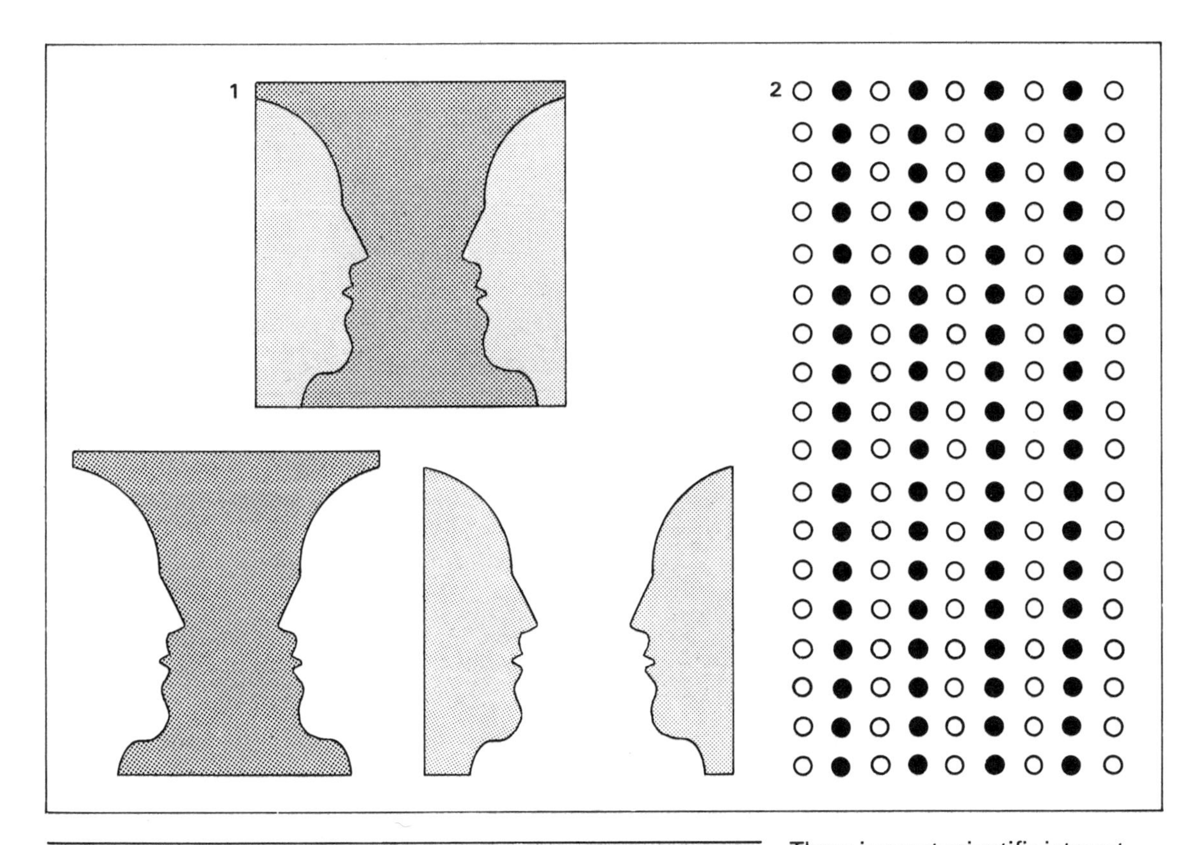

Perception

Whenever we use one of our senses, for example, when we see or hear something, at least two things happen in our bodies. First, the sense organs, the eyes or ears, respond to the stimulus and send nerve signals to the brain. Then the brain has to interpret the nerve signals or decide what they mean so that we become aware of the stimulus and receive information about it. This process is called *perception*.

Knowledge or memory can affect perception. This is why we can look at a man standing 10 metres away from us and realise that he is about the same size as the man standing only 7 metres away from us, even though the image of the first man is smaller than the second when our eye receives it. In the same way, when you hold a cup on an angle in front of you, you realise that the top of the cup is circular in shape, although your eye receives an image of an oval shape. In both these cases, we interpret what our senses tell us so that we obtain a true perception of things around us.

There is great scientific interest and research into the potential of the human mind and how it perceives the outside world. Perception and illusions are important fields of study. In the optical illusion (1), a reversal effect may occur. The top picture may be perceived as a vase but on closer investigation it looks like two faces in profile. Although the dots (2) are equally spaced, they appear as vertical columns because of their colour arrangement.

MAN AND SCIENCE

Earliest man was only concerned with survival, keeping himself alive, warm and well fed. His earliest way of observing his world was with his naked eye and the creatures in his world were described only as he saw them: that is, flying, creeping, large or small.

Language and writing

It was not until man developed language that his experiences could be shared and his knowledge could extend beyond a single lifetime. With language and then with writing he was able to pass on information. At first, this was done through paintings on cave walls. Writing as we know it developed from pictures of actual objects. The pictures were gradually made simpler until they became *symbols*. With the invention of paper, knowledge could be recorded in handwritten books, and later into printed books.

The art of writing developed from an early simple picture form of recording information and knowledge. By the use of pictures and complex hieroglyphs, early Man was able to record his thoughts and communicate with other men. These stylish hieroglyphs are Meso-American. Gradually, simpler strokes and symbols, which were quicker and easier to write down, replaced the more pictorial hieroglyphs and modern writing appeared.

Man used his brain to adapt to his environment and to invent tools and machines to improve the quality of his life. Early Man soon discovered how to spin thread (1) and weave cloth for clothing. Later, in the fifth century BC, the loom was invented for weaving cloth. The first looms consisted only of a simple frame containing several parallel beams with a shuttle to carry the thread through the warp. These were gradually improved (2).

Early knowledge

Most of man's inventions have come about because of his special needs. At first, he had only the materials which surrounded him in nature; wood and stone for tools and animal skins for clothing. Later, as people began to group together to hunt and live, they formed villages and towns. At the same time, they began to weave natural fibres together to make cloth. Man used fire to make pots and bricks from clay, and to mix native metals like copper and tin together to produce bronze. In this way, he began to make new materials that are not found in nature. Glass is another material that was first made in early times.

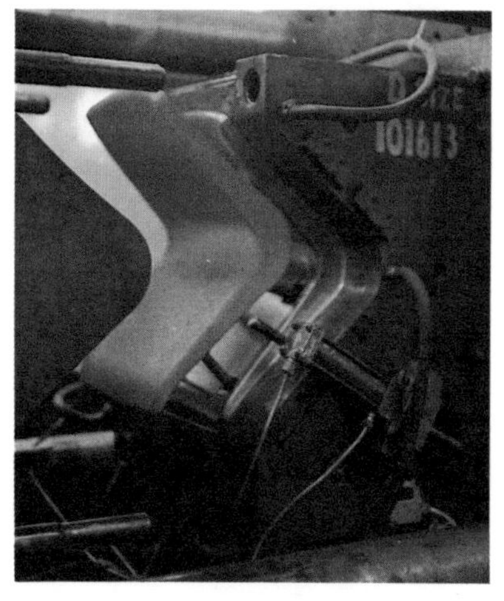

Man's scientific understanding has greatly increased in the last two centuries, which have witnessed a revolution in technology. Scientists have discovered how to make special synthetic fibres like plastic. In this injection moulding process, a polypropylene chair seat is being ejected from a mould. Plastics are now used for many industrial and household goods, ranging from furniture to food containers.

Energy and power

Man could first make use only of the power of his own muscles. Then he used the power of animals to do work he could not do himself; he domesticated animals, tamed them and brought them to live around him instead of in the wild.

Another form of energy used by early man was heat, first in his cooking fires and later to work metals. Inventions like the wheel and the plough made better use of the muscular power of animals. Later, as science advanced, man was able to use the energy of wind and water in windmills and watermills.

He discovered that burning coal gave large amounts of heat, and the development of furnaces provided enough

1

2

3

Conservation of the environment is becoming increasingly a matter of concern as industrial pollution takes its toll on our natural resources and the earth's wildlife. Scientists observe and record the effects of industrialisation and pollution on plant and animal life on the seashore (1), the countryside (2) and inland lakes and rivers (3), which may well become polluted by oil, detergents and industrial waste.

heat to make steel and to power the first modern engine, the steam engine. Railways and steamships, powered by steam, made modern transport possible. Man now had enough power available to manufacture articles in large quantities. This was the beginning of mass-production about 200 years ago.

The development of electrical power, the battery, electric motor and electric generator, gave man a source of power that led to modern communications like the telegraph and telephone and radio, television and radar.

The invention of the internal combustion engine and the discovery of sources of petroleum led on to the world of motor cars, aeroplanes and diesel powered ships.

Progress of science

The progress of technical science in this century has been as great as that of all the previous centuries put together. This is partly because new discoveries and improvements have given scientists much better instruments with which to observe and measure accurately. However, the scientific progress in earlier centuries was the foundation on which modern discoveries are built.

The main tasks of science in the future may be to help man to use his discoveries in a way that will not harm the environment or destroy it for future generations, as well as to improve things like medical science which could lead to healthier lives for human beings. Already discoveries are being made which help man to understand himself and what makes him live, as well as how life begins and renews itself through reproduction.

INDEX

Page numbers in italics refer to an illustration on that page.
Bold type refers to a heading or sub-heading.